THE

FRESH & SALT-WATER AQUARIUM

WYMAN AND SONS, PRINTERS,
GREAT QUEEN STREET, LINCOLN'S INN FIELDS,
LONDON, W.C.

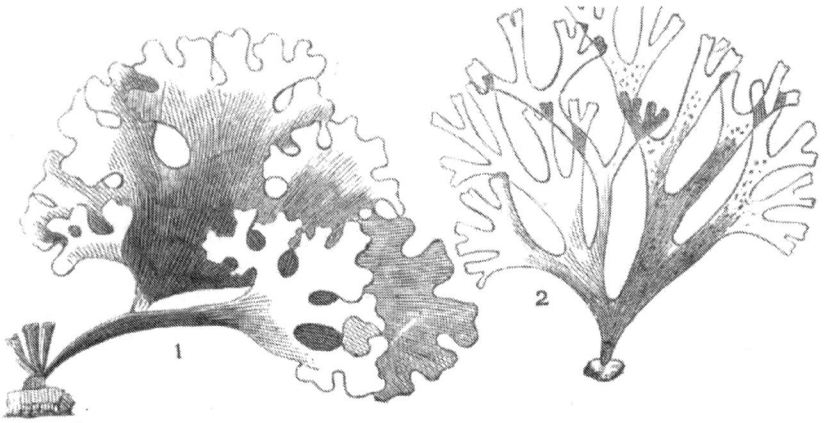

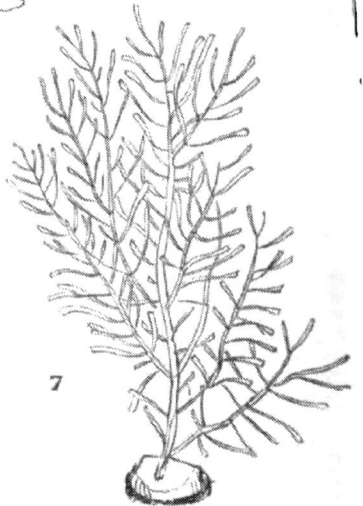

THE

RESH '& SALT-WATER

AQUARIUM

BY THE

Rev. J. G. WOOD, M.A., F.L.S., etc.

AUTHOR OF "COMMON OBJECTS OF THE COUNTRY AND SEASHORE"
"THE ILLUSTRATED NATURAL HISTORY"

WITH ILLUSTRATIONS

LONDON
GEORGE ROUTLEDGE & SONS
Broadway, Ludgate Hill
NEW YORK: 416 BROOME STREET

BY THE SAME AUTHOR.

ANECDOTES OF ANIMALS. With Illustrations by HARRISON WEIR. 3s. 6d.

MY FEATHERED FRIENDS. With Illustrations by HARRISON WEIR. 3s. 6d.

THE BOY'S OWN NATURAL HISTORY. With numerous Woodcuts by WILLIAM HARVEY. 3s. 6d.

WHITE'S NATURAL HISTORY OF SELBORNE. Edited by the Rev. J. G. WOOD. With numerous Illustrations. 3s. 6d.

THE COMMON OBJECTS OF THE SEA-SHORE. With Coloured Illustrations. 3s. 6d.

THE COMMON OBJECTS OF THE COUNTRY. With Coloured Illustrations. 3s. 6d.

THE COMMON OBJECTS OF THE MICROSCOPE. With Coloured Illustrations. 3s. 6d.

CONTENTS.

DESCRIPTION OF PLATES.

Plate II. having been spoiled in the printing is omitted. The reader will notice that no reference is made to it in the text.

FRESH AND SALT-WATER AQUARIUM.

INTRODUCTION.

IN the mythology of every land, the innate curiosity of the human mind betrays itself. Whether we take the ancient tales of Greece or Rome; whether we decipher the hieroglyphs of Egypt; whether we revel in the fanciful romances of the glowing East, or are enchained by the stern, yet not less poetical legends of the frozen North; we find that the prominent idea is the insatiable desire of possessing hidden knowledge. In some few instances this ruling passion aspires to the spirit of prophecy, but in nearly every case is contented with its present sphere, and only seeks for a more extended knowledge of the material universe, its riches and its capabilities.

Sometimes the fortunate hero is enabled to soar through the air; he ascends to the skies, and learns the secrets of the stars. Sometimes he penetrates to the central chambers of the world, anoints his eyes with magic salve, and the treasures of earth are unveiled to his cleansed vision. Sometimes he enters into alliance with the semi-spiritual beings that people the elements, and becomes a temporary partaker of their privileges. Perhaps the deities of the sea claim affinity with him, and carry him down into the ocean depths; where, with the nobles of the sea and their attendant Nereids, he holds strange festivals in submarine palaces. Perhaps the Naiads of the river become enamoured of him, decoy him beneath the waves which they rule, and keep him a very willing prisoner in their mysterious home, enchained by their beauty, and forgetful of his earthly love.

In imagination we follow their adventures, speed our course through the sky, sink into the earth, or traverse the waters, and picture to ourselves the wonders which such privileges would reveal to us.

> " Methought, I saw a thousand fearful wrecks;
> A thousand men that fishes gnawed upon;
> Wedges of gold, great anchors, heaps of pearl,
> Inestimable stones, unvalued jewels,
> All scattered in the bottom of the sea."

Specially does imagination revel in the depths of ocean, and disport itself among the caves of ocean, paved with gems and pearls, hung with living tapestries of sponge and seaweed, and peopled with strange, weird shapes, that crawled upon its sides, or bright glittering fishes, that darted through its portals.

Thanks to modern inventions, many of these hidden wonders are now disclosed, and the curious inquirer can see these beings without the assistance of either Nereid or Naiad. Some of these wonders will be described in the following pages, and those who will endeavour to search for themselves will find that not even the vivid pages of fairy lore could picture beings of stranger aspect, or endow them with more singular power, than may be seen in a few gallons of water while we sit at ease in our own rooms. We need not take the trouble to descend in a diving-bell, nor fit on our head the diver's enormous helmet; for we can find in a small vessel of salt or fresh water enough to give us occupation for many a long year, and to disclose to us many of the secrets of nature. In the following pages I shall endeavour to show how any one can penetrate into the wonders of the deep, and can, in the seclusion of his own room, study the manners and customs of the inhabitants of the waters.

The history of aquaria is quite recent, but in the few years of its existence it displays many of the characteristics of more important histories, and has its origin, its rise, its decadence, and its renovation. Some years ago, a complete aquarium mania ran through the country. Every one must needs have an aquarium, either of sea or fresh water, the former being preferred.

There were grand etymological discourses in the learned papers respecting the correct name which ought to be given to it. Some called it vivarium, but were met by objectors who said the Zoological Garden was equally a vivarium, and so was a dog-kennel or a stable. In order to meet the difficulty, they proposed the word aqua-vivarium—a word which certainly had the advantage of being correct, but the disadvantage of being complicated. Then came others who preferred the name aquarium, and straightway this name was adopted by common consent. It is true that exact linguists rejected the word, citing the Latin dictionary, which states that *aquarius* was either a water-bailiff, a water-man, or "the man who carries the watering-pot" in the Zodiac. Still aquarium is a simple and easy word, and entirely superseded aqua-vivarium, just as in a later year the word telegram superseded telegrapheme.

The fashionable lady had magnificent plate-glass aquaria in her drawing-room, and the schoolboy managed to keep an aquarium of lesser pretensions in his study. The odd corners of newspapers were filled with notes on aquaria, and a multitude of shops were opened for the simple purpose of supplying aquaria and their contents. The feeling, however, was like a hothouse plant, very luxuriant under artificial conditions, but failing when deprived of external assistance.

Perhaps the beautiful plate-glass aquarium fell to pieces, discharged several gallons of sea-water over the fashionable carpet, and covered the fashionable furniture with sea-anemones, crabs, prawns, and other inhabitants of the waters. Or some of the inmates died, and the owner was too careless to remove them. Consequently they were left in the holes to which they had retreated, and in a few days they avenged themselves for the neglect by rendering the water so fetid that no one with ordinary sensibility could remain in the room. The schoolboy was very careful of his aquarium for a time, but in a month or two became tired of the constant attention required for its maintenance, and so gave it up.

So, in due course of time, nine out of every ten aquaria were abandoned; many of the shops were given up, because there was no longer any custom; and to all appearance the aquarium fever had run its course, never again to appear, like hundreds of similar epidemics.

But there was one element of strength in the aquarium possessed by none of the others. This was the study of Nature in one of her hitherto unstudied phases. Those who merely treated the aquarium as a toy soon became tired of it, and cast it away accordingly, but those who saw its real capability became more enamoured of it daily. Now, therefore, the number of aquaria is not nearly so great as was the case some years ago, but those that are in active existence are properly tended, and the teachings carefully learned. As I hope that many of my readers will desire to establish and to maintain aquaria, both of fresh and salt-water, I will give a few hints as to the stocking and managing them.

Now, in the first place, the very name of AQUARIUM terrifies many, especially those who are unversed in its easily-solved mysteries. The very name seems somewhat pretentious, and the popular idea of the aquarium is quite consonant with such an idea. The word aquarium suggests an expensive structure of plate-glass, shining metal, and elaborate rockwork, tenanted with rare actiniæ, strangely-shaped crustacea, and gorgeous fishes, and decorated with seaweed of light green, dark purple, and rich scarlet.

Of course, it is perfectly possible to have such structures, and to fill them with such inhabitants; but for all practical purposes there is not the least necessity for either the elaborate vessel or the costly inmate. After an experience of some twelve years, I am disposed to multiply the vessels rather than to expend money in animals; and have found that the simpler a vessel can be, the better it is for an aquarium, provided it fulfils certain conditions.

There are two forms of vessels which are generally offered to the public, and both of them are about as bad as bad can be. The one is a deep, oblong vessel made of plate-glass, with a metal bed and metal framework. Set a pack of playing-cards on its edge, and there is the precise shape of the aquarium generally sold in ordinary shops. This being rather expensive, many persons demur at its price, and are then offered the ordinary gold-fish globe, or an improvement thereon—a vessel that looks like a glass bell standing on its head, and fixed into a wooden pedestal. I shall now proceed to show why these forms are entirely wrong; and then describe a variety of shapes which are well adapted to the purpose, and which are equally easy to be obtained and cheaply to be purchased.

In the first place, all glass vessels have one radical error : they admit too much light; and, as a general rule, the creatures which inhabit the water are not lovers of light. Many of them lurk unseen in holes, or under stones, or among roots ; and even those which do show themselves boldly are always glad of darkened places of refuge, whereto they can fly when alarmed. Vessels wholly made of glass afford no such privileges, and therefore the inhabitants are often sadly inconvenienced, and in 'many cases die, simply because they are worn out by want of repose.

Moreover, as if the poor animals were not sufficiently worried, the aquarium is often put on a table or stand, and placed exactly in front of a window, so as to have as much light as possible upon it, and to have the water warmed by the direct rays of the sun, until the miserable inmates are half-killed with the heat and totally dazzled by the light. Such an excess of light is injurious in another sense, especially in marine aquaria, because it encourages the growth of conferva and seaweed to such an extent that they become a positive nuisance, instead of a useful accessory.

The form of the ordinary-shaped tanks is quite wrong. They certainly hold a large quantity of water, and therefore are thought fit to contain a large number of inmates. But it must be remembered that fishes, and other inhabitants of the waves, exhaust the water by their respiration quite as much as we exhaust air.

If you put a quantity of fishes into a small and deep vessel full of water, some will die in a very short time, and only a few hours will elapse before all will have perished. It is therefore as absolutely necessary that the abstracted oxygen should be restored to the exhausted water, as that the atmosphere of a crowded room should be replenished with fresh air. In consequence of this necessity, various plans for aëration were employed by aquarium-keepers. Ingenious pumps were attached to the tanks, by means of which streams of air were forced through the water. Some persons employed syringes, filled them with water, and squirted the water into the tank with such force as to carry a quantity of air among the inhabitants of the aquarium. Others were content with taking up some of the water and letting it fall back with a splash, so as to produce the same result.

Now, all these proceedings were absolutely necessary, on account of the shape of the tank, and the stillness of the water. Many marine creatures, such as certain crustacea, molluscs, and sea-anemones, live close to the shore, and are accustomed, not only to be left dry during extreme low water, but to have the spray dashing about them twice a day, as the advancing tide breaks over the rocks or sand. But the very form of the common oblong tank opposes itself to both those conditions. It is so deep that a perfect stillness reigns, and presents so small a surface to the air that there is no chance of oxygenizing the water except by artificial means. Water absorbs the oxygen of the air with wonderful rapidity, and if a sufficient surface be exposed, it will absorb enough to supply the wants of respiration for a goodly number of inhabitants.

Were it not for this fact, the fishes in a pond would soon die for want of oxygen.

It will now be seen that an aquarium which is to fulfil, as far as possible, the same conditions as the river, the pond, or the sea, ought to be as wide as possible, so as to present a large superficies of water to the air. Moreover, it must not be made of a transparent material, such as glass, but its sides ought to be opaque, except in front ; and the front should not be turned towards the window. Should the reader happen to possess one of these ordinary tanks, he can vastly improve it by covering the back and the ends with thick pasteboard, so that the light is shut out, and the pasteboard can easily be removed for the purpose of inspecting the interior of the tank.

But there is no need whatever for a complicated glass tank, which is so deep that the owner finds great difficulty in getting at the various objects, and is too heavy to be moved, and occasionally apt to worry its owner by a sudden disposition to leak. Any kind of tub or pan will do for an aquarium, provided that the owner cares more for the inmates than the appearance of their dwelling. I have now at my side a common earthenware pan, eighteen inches wide, and three deep, in which are flourishing half a dozen sea-anemones, two kinds of sea-weed, and a number of purpura and other common shells.

A crab lived in it for a considerable time, and would probably have been alive now but for a singular misfortune. One evening, before the lamps were lighted, I was coming up the stairs to my room—it is situated at the top of the house ; and upon the third stair from the front door I trod upon something, and crushed it. On bringing a light and looking to see what strange object could have been on the stairs, I was equally surprised and sorry to find that it was my little crab. It was all but dead, and recovery was hopeless. The fact was, that I had forgotten to cover the pan, and the crab must have clambered up the side, and, hitching itself on one of the shells, got out of the pan and traversed the room. But how the creature contrived to get out at the door of my room, and make its way down two flights of stairs and along a landing, is a mystery which I cannot solve.

I was the more sorry because the creature had learned to overcome its first dread of a human being, and, instead of scuttling under a stone when I approached, remained quite unconcernedly on the top of an ulva-covered flint, or crawled leisurely over the bottom of the pan. At first it was sadly discomfited at its inability to burrow; the yellow colour of the pan evidently giving it the idea that it was upon its accustomed sand; but it soon gave up the attempt, and seemed quite reconciled to its prison.

In this primitive tank there are some flints, with a little ulva and enteromorpha adhering to them. I had some with specimens of the ocean barnacle; but as the creatures declined to live, the stones were removed. As to the anemones, they have chosen to fix themselves so tightly in an old pomade pot in which they were conveyed to my house, that I have not disturbed them.

Of course, in a vessel of this kind, the loss of water by evaporation is very rapid, and must be repaired by constant supplies of fresh water. There is not the least difficulty in adding the needful water, nor need it be distilled, as is stated by some aquarium-keepers. I simply use soft water, taking care that it is clear, and pour it into the pan without any precaution. I do not pour it upon the anemones, because the fresh water might fill the jar in which they have stationed themselves, and so damage them seriously. Otherwise, I just pour in the water, and let it mix itself as it likes, without stirring it or doing any thing to disturb the inmates.

Once or twice, when I have been away from home for a day or two, the evaporation has been equal to nearly one-third of its amount, and the remaining fluid formed, in consequence, an exceedingly concentrated solution of marine salts. The inhabitants did not like the state of things at all. The shells were all out of the water, and the anemones much contracted, although the tips of their tentacles protruded. When the fresh water was added, they withdrew the tentacles entirely, and made themselves as flat as they could; but in an hour they had recovered themselves, and were waving their long tentacles in all their beauty.

If the young observer is lodging at the sea-side, let me advise him to set up—not one large aquarium, but a series of pans, previously taking care to propitiate the landlady, who is sure to offer very forcible objections to such articles. Get a glazier to cut glass covers for your pans, or the inmates will escape, and the pans be filled with dust.

We will now proceed to a short description of the various vegetable and animal inhabitants of an aquarium, and will begin with the simplest; namely, the Seaweed.

There are many species of seaweed, which are either pretty in colour, graceful in form, or imposing in appearance, which can never be kept in an aquarium. Not only will they die with great rapidity, but they cause an ugly appearance in the tank, giving out slime and other unpleasant substances, and often producing an odour peculiarly abominable, which cannot be easily described, but which, when once smelt, is never forgotten.

As a general rule, the young aquarium-keeper may reject almost every seaweed that he finds, and to this rule there are but a very few exceptions. He may think it rather hard to be obliged to do so, because some of them, especially those belonging to the red division, are so pretty that they are for a time extremely ornamental, while others are so elegant or so curious in shape, that it is, at first, very hard to resist the temptation of placing them in the tank. Such, for example, is the Delesseria, whose fronds may often be found on the shore after a gale, having been torn from their attachments by the waves. It is so elegant in shape, and so lovely in colour, that even an experienced aquarium-keeper can seldom pass it, as it lies on the shore, without picking it up and admiring its exquisite fronds, so delicate in structure, and so leaf-like in form. Nothing looks better in an aquarium for a day or two; but then its beauty fades, the frond becomes paler in spots, then is ragged at the edge, and finally breaks up into fragments. So it is with the curious corallines, with their strange chalky skeleton; and so it is with the whole of the red seaweeds, in spite of their enticing beauty.

As to the brown varieties, many of them are so large that they could not be placed in any ordinary tank, but there are some which are small enough to find space, and curious enough to induce a beginner to place them in the aquarium. Such, for example, is the strange-looking Padina, whose fan-like fronds, with their regular dark bands, are not very frequently found on our coasts, except those in the extreme south.

Some years ago, however, I observed a large colony of the Padina growing upon a ridge of rocks running seaward from Foreness Point, at Margate. Every visitor to this place knows the rocks, dark with their heavy covering of seaweeds, that project into the sea, and are left bare at low water. Through these rocks run, at various intervals, certain channels which are free from the bladder-wrack, and other large algæ, with which the rocks are so thickly clothed, and which serve to conduct the waters of the receding tide back to the sea.

One day, at very low tide, when wading along one of these channels, I saw a long ridge of some kind of algæ, different from those species which were most prevalent, but, owing to the ripple caused by a sharp breeze, could not discover what it was. On gathering it I was surprised to find that it was the *Padina pavonia* itself—just the very last species I should have expected to find at Margate. One might hope to discover many of the more hardy species, but to find an alga which is mostly confined to the extreme south, off the Margate shore, which lies open to the north wind and gets the full benefit of it, was a circumstance which could hardly be expected.

CHAPTER I.

ALGÆ, OR SEAWEEDS.

THE green varieties of algæ are most easily kept, and the very best of them is, fortunately, the most plentiful. This is the *Ulva latissima* (Plate I. fig. 4), whose broad green, ribbon-like fronds are to be found in abundance. It is popularly known as Green Laver, because, when gathered and stewed down slowly, it is reduced to a gelatinous substance which is called laver. The laver is said to be extremely nutritious, like the Carrageen or Irish moss, which also grows in abundance on our coasts, and by some persons is very much liked. An allied species, which will be presently mentioned, is employed in the same manner. The green laver is sometimes called the sea-lettuce, on account of the shape and colour of its fronds.

There is no better alga for an aquarium than the green laver. The reader will remember some remarks upon the mode of supplying oxygen to the water, and the trouble which is sometimes involved in performing that operation. Now, the ulva saves much of this trouble, and by its own respiration takes from the water those elements which the water does not require, and supplies those which it needs. If a frond of ulva be placed in some sea-water, and set in a tolerably light spot, a most beautiful appearance will be presented. The entire surface of the frond will be covered with minute bubbles of air, looking as if diamond dust had been scattered over it. By degrees these bubbles enlarge, then coalesce together, and finally are detached from the plant and float to the surface. If the contents of these bubbles be gathered together and examined, they will be found to consist chiefly of oxygen gas, just the very element of which the animal inhabitants deprive the water by their respiration. The plant, in its turn, requires the elements exhaled by the animals, and so the circle of nature completes itself.

If the aquarium-keeper would like to test the quality of this gas himself, he can do so in a very simple manner. Let a bottle be filled with water, and then placed, mouth downwards, in the aquarium. The mouth of the bottle should then be held over the leaves, and the bubbles touched with a small stick or wire. They will then become detached, and will rise into the bottle, displacing their own bulk of the water. When all the bubbles are collected, the stopper is placed in the bottle, while it is still under water, and the bottle can then be removed with its contents.

There is no necessity to have a large quantity of these algæ, and, indeed, there is no absolute necessity to have any algæ at all, if the vessel is wide and shallow, as has already been described. Moreover, if the conditions should be favourable, the multitudinous spores will settle themselves on the stones or sides of the tank, and will cover them with a soft coating of minute algæ, which will serve the double purpose of looking pretty and aiding in regenerating the water. The more light that is admitted, the faster will these spores develop themselves, and it is often necessary to keep a sponge fixed to a handle for the purpose of keeping the glass front of the aquarium clear from these tiny settlers.

Another useful alga is the Purple Laver (*Porphyra laciniata*). In form this species bears a close resemblance to the green laver; but it is very different in colour, being darkish purple instead of green. In some parts of this country it is much eaten, being cooked in the same manner as the green laver. In Ireland it forms a favourite article of food, and in many places is always cooked and brought to table in a silver saucepan. When cooked, it is popularly called "sloke."

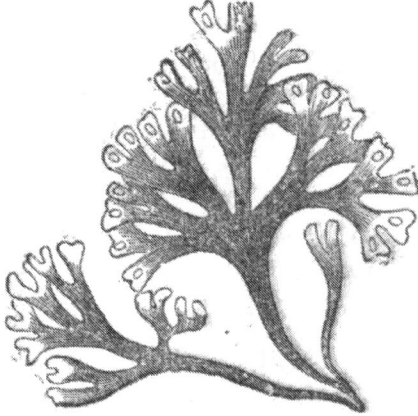

Chondrus crispus.

As we are on this ground, I may mention that two more edible seaweeds are plentiful on our coasts. One is the Carrageen (*Chondrus crispus*),* which is often known under the name of Irish moss, and which, when reduced by boiling to a gelatinous consistency, is mixed with tea and coffee, or sometimes eaten by itself as jelly. This alga is very common, and any quantity may be gathered in an hour. It is extremely variable in colour, the hues differing according to the locality and depth of water. Generally it is dark green, with a tinge of yellow near the edges; but, in deep water, the yellow hue is not developed, and, in consequence, the frond assumes a purplish tint. Light seems to produce the yellow colour; for in specimens that grow in the pools near high-water mark, and which,

* See also Plate I. fig. 2.

in consequence, are much exposed to the direct rays of the sun, the colour is greenish-yellow, the latter hue predominating.

When washed in fresh water and well dried, it becomes almost horny in texture; and when macerated in hot water, it dissolves like gelatine. This gelatinous substance is, by the way, admirably adapted for fastening dried seaweed to paper. In most marine algæ there is a sufficient amount of this vegetable glue to make them adhere to paper as firmly as if paper and alga were one substance. But there are several species that will not adhere without external aid. Various materials have been employed for the purpose; such as gum, isinglass, and liquid glue. But there is nothing which really answers the purpose so well as gelatine made from the Irish moss. This species can be kept in an aquarium, if care be taken. But it is large, occupies much space, and is certainly not rare enough, nor handsome enough, to be worth the trouble.

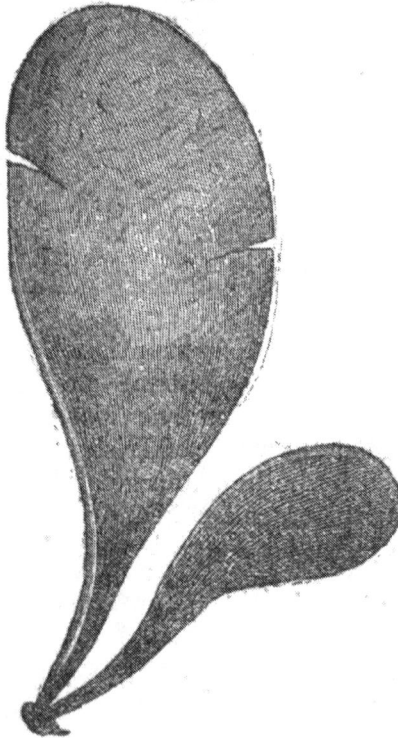

Iridea edulis.

The last species of edible seaweed which will be mentioned is the Dulse or Dillosk, a species which is found on many of our coasts, and which is very plentiful in some localities. Its form is given in the illustration, and its colour is a deep, dark red. It derives its name of Iridea from the iridescent hues which play on its surface when submerged, and which change with every movement of the plant or of the spectator. The dillosk is a favourite article of food in many places, especially on some of the coasts of Ireland, where it is eaten both raw and cooked. Some persons are so extravagantly fond of it that they always have it served up for breakfast. The taste, however, seems to be an acquired one, and it is certainly one which I never could acquire, though I have given it a fair trial. Perhaps the taste for dillosk is like that for olives, and comes naturally to some persons, while others find themselves unable to perceive any gratification in eating either the one or the other.

Among the green seaweeds that may be kept in

Cladophora arcta.

in aquarium may be mentioned the very prolific

species which is scientifically termed *Cladophora arcta*. This is by no means a pretty species when taken from the water, though it is handsome enough when submerged.* But, to the seaside observer, it is one of the treasure-houses of the deep. Should the reader of these pages like to go to the seaside and examine for himself, let me recommend one excellent plan.

Mark out a small tract of shore, say fifty yards wide, and work it thoroughly. Off with such impediments as respectable shoes and stockings; but put on an old pair of shoes, with tolerably stout soles, or you will find that wading is not so pleasant a process as it looks. There are sharp stones; there are limpet shells, which cut the feet whichever way they may be lying; there are the acorn barnacles, which render the walking as agreeable as the celebrated journey with peas (unboiled) in the shoes; there are mussels, which make the unlucky wader think that he is walking over a road composed of broken bottles; there are sharp splinters of wood, and, in fact, everything that makes the wader look to his footsteps, and not to the inhabitants of the sea. The area being rigidly marked out, let every square foot be carefully examined, beginning as soon as the receding tide will allow the inhabitants of the water to be seen, and continuing until the waves begin to return on their endless track. Look into every crevice, examine each projecting angle of rock, turn over each loose stone, lift up each bunch of seaweed, and pick up every shell that you come across. Do not be disheartened at the slowness of the process, for it is the only method of working a locality thoroughly. At first it is rather tedious work, simply because

* Another species, *Cladophora pellucida*, is shown at Plate I. fig. 3.

the eye is not accustomed to notice those minute details on which the whole science depends. Unless trained in this careful manner, the eye is sure to miss little points which speak volumes to the experienced naturalist, but which a novice can hardly see at all, and, if he does see, cannot understand. In a few days the improvement will be perceptible, even to the observer himself, and he will find that he is taking notice of minutiæ which before would have escaped observation, and which, although they appeared trifling, are as valuable to him as a nibbled grass-blade, or a leaf turned the wrong way, to an Indian tracker.

The Cladophora is just one of those very objects that well repay a close examination. You will neither have time for useful examination while still wading, nor will you be able to stoop for so lengthened a period as would be required. Look out for a good thick bunch of the Cladophora, pull it up gently at the root, and transfer it to a pail or similar vessel. Do so until the pail is full, and postpone the examination until you reach home. There have a large empty basin, and a plentiful supply of sea-water.

Pour a quart or so of water into the basin; pick off a small tuft of the Cladophora, and put it into the water. Although it clings so tightly together while in the air, no sooner does the alga find itself in water, than it spreads itself out, and all the little branchlets are supported by the liquid. Now, shake each tuft in the water, and separate the fibres, and you will be tolerably sure to find among the green foliage some of the smaller animals which inhabit the ocean, star-fishes and crustacea of different kinds being the most plentiful.

The last of the seaweeds that will be mentioned is the *Bryopsis plumosa* (Plate I, fig. 6), a singu-

c 2

·larly elegant little species, for which I very much
regret there is no popular name. Its colour is
very light green, and, when growing, it has the
appearance of green feathers set on slender stalks.
Being an annual plant, it can only be found during
certain seasons of the year, such as summer and
autumn. In proportion to its size, it has a large
amount of gelatinous substance, and adheres well
to paper.

CHAPTER II.

FISHES.

HAVING placed the seaweed in an aquarium, the next step is to stock it with living inhabitants. Of the dwellers in the sea, the fish take the first place, and will therefore be mentioned first.

The number of species found on our coasts is so great that only a very few can be mentioned in these pages. As the size of some of these unfits them for residence in the limited space afforded by an aquarium, I shall therefore only mention a few of those which have actually been kept in an aquarium, and which have some peculiarity that is worthy of notice.

Perhaps the reader may ask how the fish in question are to be caught, inasmuch as the sea is rather a wide place, and means of capture are limited.

It is evident that the hook is not a desirable instrument, and that it must therefore be laid aside whenever the fish are wanted in a perfect state. The net, in one or other of its forms, is the only implement that is really serviceable; and this must be modified according to the locality, and the particular fish which is to be captured. Some species, for example, are seldom, if ever, seen in deep water, preferring to hang about the rocky shores, to ferquent the sandy coasts, or to haunt those spots where the zostera waves its long narrow blades, like a submarine hayfield just ready to be mown.

Some, on the contrary, never trust themselves to the shallow water, and are only to be captured at a depth of many fathoms. Most of these, however, may be found within a mile or so of the shore, according to the slope of the sea-bed, the consequent depth of the water, and the particular sorts of seaweed which grow in the locality, and which furnish retirement to the marine animals on which they feed.

Of all the varieties of nets, those which are most useful for the owner of an aquarium are the hand-net, the trawl, the dredge, and the keer-drag. The first of these nets is really very useful, and will enable a skilful handler to capture many kinds of little fishes that are left in the rock-pools when the tide has receded : gobies, gunnells, pipe-fish, and the like, being the usual inhabitants of such pools. The trawl, in places where it can be used, is simply invaluable, as it sweeps into its treacherous mouth all kinds of animals, including many species of fish, and creates a positive embarrassment of riches to the trawler. The dredge, though it is mostly employed in procuring molluscs and crustacea, is extremely useful in the capture of fishes, which, in spite of its small size, are frequently found in it. The keer-drag is a modification of the dredge, but has a very long net, with an arrangement which allows the fish to enter, but prevents them from retreating.

As to small flat-fish, such as dabs, flounders, &c., nothing is better than the common shrimp-net. If the reader does not object to getting wet, he cannot do better than walk through the sea with a shrimp-catcher, and give him some remuneration for taking a choice of the contents of his net. It is useless to ask him to take the shrimps and prawns, and put all the other creatures in a separate net. He will not do it. He has such an inveterate habit of throwing away everything that is neither a shrimp nor a prawn, that he mechanically empties into the sea the very creatures which are most required. In most cases, he does not recognize a vast number of animals to be animals at all, but classes them with seaweed and stones, and other "rubbish."

THE first fish in our list is the common Basse (*Labrax Lupus*), which is closely allied to the fresh-water perch, and which, in fact, looks very like an ordinary perch that has contrived to escape from the river into the sea. On some of our coasts the Basse is called the Sea Dace. As, moreover, this fish is somewhat tolerant of fresh, or, at all events, of brackish water, the two species might easily be confounded together. It has even been transferred from the sea to an inland lake, and has been rather improved than deteriorated by the change. This hardiness of nature causes it to be an admirable tenant of an aquarium, where the daily conditions of life are so unlike those to which the inmates are accustomed when in the enjoyment of freedom.

It is essentially a shore fish, finding all its food within a very short distance of the land, and being nearly as bold as its fluviatile relative. It is particularly fond of the Onisci, or Sea-slaters; and when the wind is high, and the waves are dashing on the shore, the Basse ventures quite close to the rocks, for the purpose of picking up the unfortunate slaters, as they are washed out of the crevices in which they hide themselves by day. The Basse may be caught like the Perch with a rod and line, and bites most freely during the last portion of the flood-tide.

A long jetty like that of Margate is a good spot wherein to angle for the Basse as well as for other shore-loving fish ; they seeming to be attracted by the smaller marine animals that haunt the piles on which the structure is supported, or by the *débris* that is usually thrown in the water. Sometimes it is taken with a very simple kind of net, resembling a seine on a very small scale. This net is taken in the water by two men, who wade as far as they can, and then separate, each taking an end of the net. When they have stretched it, they walk slowly ashore, bringing with them the fish entangled in the net. Smelts are taken in the same manner in the Medway and other localities which they favour with their presence.

The colour of the Basse is dark-blue above, and silver-white below. The dark bars which are so conspicuous on the Perch are but faint in the Basse, and the fins lack the bright scarlet which distinguishes those of the Perch. The body is not so deep in proportion to its length as that of the Perch, and the gill-cover is armed with two pointed projections, directed towards the tail. In an aquarium it is a brisk and lively fish, and small specimens ought to be selected.

Next comes the wide-headed, angular, armour-clad fish, the Pogge (*Aspidophorus Europœus*). This little fish has a great variety of names, as is likely to be the case on account of its extraordinary appearance and its prevalence on our coasts. In some places it is called the Lyrie, in others the Sea-poacher, in others the Armed Bull-head, in others the Pluck, and in the North it goes by the name of Noble.

Were it only for the singularity of its aspect, the

Pogge is a worthy inhabitant of an aquarium. I never kept it myself, but Mr. Hillier tells me that he has kept specimens for a considerable time in his aquarium, which is of moderate dimensions— three feet long by eighteen inches wide and fifteen deep. It is one of the fishes that can be obtained by accompanying a shrimper, as it is frequently taken in the net together with the shrimps and prawns, and invariably thrown back into the sea.

It is impossible to mistake the Pogge for any other fish, as it is at once known by the vast number of projecting points with which the body is covered. It has sharp points to the dorsal fin, the head is all angular and pointed, and eight rows of pointed tubercles or scaly plates run along the body, covering it with a suit of plate armour, and giving it an octagonal shape. The colour of this odd little fish is brown above, with four broad dark-brown bands; the under part is very white, and the large fan-like pectoral fins have a brown bar across the middle.

One of the most easily kept of sea-fishes is the common Grey Mullet (*Mugil capito*). This fish is very plentiful on our coasts, and can be taken without difficulty. Like the Basse, it is equally capable of inhabiting fresh and salt water, and is in the habit of ascending rivers in search of food. It can then be captured with a fly, and gives good sport, as it is an active and even intelligent fish, accustomed to freedom, and not being able to endure the idea of captivity.

If the Grey Mullet be placed in an aquarium, the owner must take care that a cover be placed on the top, or the fish will assuredly leap out of the vessel and be lost. Mr. Couch states that the fish is so impatient of confinement, that even when mullets have been imprisoned in a large pool of twenty acres in extent, they have been known to leap fairly on the bank, in their endeavours to regain the open sea. If caught in a net, the Mullet is sure to use every method of escaping, and generally prefers to leap over the top of the net before it becomes entangled in the meshes. If one succeeds, the rest are sure to follow, like a flock of sheep leaping over a hedge; and unless the net can be raised at once, the whole of the inmates will escape.

Naturally the Grey Mullet has a liking for brackish water, and during the earlier portion of its life hangs about the mouths of rivers. Sometimes it has been known, even when full-grown, to penetrate full twenty miles inland. I believe, however, that it always returns seaward with the tide. Some time ago, the same gentleman who experimented upon the Basse, transferred a quantity of young mullets to the pond, and found that they throve well, grew rapidly, and were heavier in proportion to their length than those taken out of the sea.

THERE is a very common little fish, very eel-like in form, and which is often mistaken for an eel, as it goes undulating through the water. This is the Spotted Gunnell (*Gunnellus vulgaris*), sometimes called the Butter-fish, on account of the slippery secretion with which its body is covered. It may be found in the rock-pools after the tide has receded, and is not very easy to catch, as it has a way of darting into crevices which scarcely seem able to hold a fish of half its size. The net cannot reach it in these hiding-places, and if it be grasped in the hand, which is not often the case, owing to its agility, it slips through the fingers just as the well-oiled thief of India eludes the grasp of the pursuer. I have often amused myself and some friends by finding a gunnell in a rock-pool and trying to capture it with the hands alone. The chase was always a long one, and it was very amusing to see the agile creature wriggle its way from one hiding-place to another, and then, even when caught in the hand, slip quietly through the fingers, and leave its captor to begin the chase afresh.

The colour of this odd little fish is rather variable, but is always brown of some kind, more or less dappled. The name of Spotted Gunnell is derived from a row of black spots edged in front and behind with white, which are placed along the base of the dorsal fin. This fin is very narrow, and extends along the whole of the back. The scales are extremely small, and on account of the mucous secretion which covers them, are scarcely visible unless very carefully searched for. It is a very hardy species, and can endure a long absence from the water, if it be laid on damp seaweed. Its average length is six inches.

THE Blennies and the Gobies all belong to the same family as the Gunnell, and are even better adapted for the aquarium. Montagu's Blenny (*Blennius Montagui*) is, according to Mr. Hillier's experiments, the very best of all fishes for the aquarium. It may be found in rock-pools like the Gunnell, and can easily be captured by means of a hand-net and a stick, the latter being used for the purpose of driving the little fish from its hiding-place, and directing it towards the mouth of the net. It is a prettily-coloured fish, dark green above, with pale blue spots, and white below. The broad pectoral fins are spotted with orange.

The Gobies are easily caught and easily kept, and of these I have selected the Black Goby (*Gobius niger*) for this paper. The chief peculiarity of the Gobies lies in the manner in which their ventral fins are united at their edges, so as to form a slightly hollowed disc. This disc is in form very much like a boy's sucker, and is used for the same purpose. A similar structure is seen in the common Lump-sucker, only in this fish the pectoral fins are also merged into the sucker, and the adhesive power is so great that it has been known to raise a weight of sixteen or eighteen pounds.

In the aquarium, the Gobies will mostly exhibit the powers of the sucker, darting about with great velocity, and sticking suddenly upon a stone, as if arrested by some external power. I do not remember seeing them adhere to the glass sides of my aquarium, though they probably might have done so had there been no smooth stones. After a little while the Gobies will become quite reconciled to their lot. At first, they are extremely nervous, darting wildly about at the least movement, and reminding the observer of the frantic struggles of newly-caught sparrows. But in a few days they become accustomed to captivity, and, if properly managed, will come to the surface of the water and take food almost out of the fingers. A slight paddling in the water with the end of the finger suffices to attract them.

Although small, and possessing teeth of proportionate size to their bodies, they are exceedingly voracious, and, if they were as large, would probably be as destructive as the iron-jawed Wolf-fish,

which belongs to the same family. The Goby is
very active in pouncing upon prey, whether living
or dead; and when it has seized the coveted object,
it returns to some neighbouring hiding-place, and
there eats it if dead, or carries out the struggle if
living. The reader will remember that the Pike
has a similar habit, and so have many other predaceous fish. It is but a little fish, averaging four
inches in length. As its name imports, it is of
very dark colour, except in a portion of the under
surface, which is whitish-grey. Like the Gunnell,
it is covered with a mucous secretion, which renders
it exceedingly slippery to the touch.

THERE are one or two curious little fishes inhabiting the waters of the English coast, which are
sometimes called Dragonets, and sometimes Skulpins. They are peculiar in appearance, and cannot
be mistaken for other fish. The most curious of
them, the Yellow Skulpin, or Gemmeous Dragonet
(*Callionymus Lyra*), is notable for the enormous
length of the first dorsal ray, which, when the fish
is full-grown, reaches as far as the tail, and sweeps
over the back in a bold curve. It is a very bril-
liantly-coloured fish, the general hue being bright
yellow, with spots and stripes of blue on the head
and sides, and white on the under surface. In
allusion to the peculiar hue, the Scotch fishermen
call it the Gowdie.

This beautiful creature is, however, the adult
male, and is not very plentiful; but the females
and immature males are much more common, and
not nearly so richly decked. These creatures are
called Common Skulpins, and are sometimes termed
Sordid Dragonets or Foxes; the former of these
names alluding to the comparatively dingy hues.

and the latter to the reddish-brown colour of the back, which is of very similar hue to the fox's coat. These fishes were long thought to be separate species, and were known by separate scientific names; but Dr. Günther, of the British Museum, has satisfactorily settled the question of their identity.

CHAPTER III.

FISHES—*continued.*

D R. GÜNTHER very well observes, that no female with an elongated spine and yellow covering has ever been discovered, and that the slight variation in the habits of the Gemmeous and Sordid Dragonets is caused, not by any distinction of species, but by difference of age. The mature fish is quick in its habits, and likes moderately deep water, while the immature specimens are brisk, active, and lively, darting off at any passing object which they may think suitable for food, and then returning to their resting-place, just like fly-catchers pursuing their prey.

The Dragonets are mostly captured in trawl nets, and are frequently taken by shrimpers, though the capture of an adult male in a shrimp-net would be rather a rare occurrence. If the reader can secure a good specimen, let him make much of it, and not fail to preserve it when it succumbs to the fate that at one time or another befalls all inhabitants of an aquarium.

THERE is a small group of fishes which are called, from their structure, Pipe, or Bill Fishes. They are among the oddest of the finny race, and are well worth a careful examination. Fortunately, they are very plentiful, so that there is no difficulty in obtaining specimens for investigation.

In this group there exist so many peculiarities

that it is difficult to describe them in their proper order. We will, however, begin with the structure from which they derive their name. They are among fish what the Snipe and Woodcock are to birds, their jaws being greatly lengthened and attenuated. But they differ from the Snipe in the fact that the jaws are united throughout their whole length, and have but a very sm ll mouth at the extremity. This structure of the jaws and mouth is almost exactly similar to that which is found in the Echidna, or Spiny Anteater, of Australia, and it is impossible to compare the two creatures together without seeing that the same principle has been carried out in a denizen of the land and a dweller in the water.

The gills of the Pipe-fish are not in the least like those of the fishes which have been already described, and which are familiar to us by means of the fishmongers' shops. Instead of the scarlet fringes which deck the branchial arches, the Pipe Fishes have a number of little round tufts set closely together.

The body is long and snake-like, and in one species, which is scarce, and not likely to be taken near the shore, the body is not thicker than a goose quill, though a foot or fourteen inches in length, and tapers away almost to nothing at the tail. The end of the tail is prehensile, and is used exactly as are the tails of the various prehensile mammalia which inhabit trees, and use their tails as a means whereby they can suspend themselves from the branches, or more firmly maintain their position, whereas the Pipe-fishes use them in order to grasp the seaweed and anchor themselves in safety while the restless tide is passing by them.

Pipe-fishes are very interesting inhabitants of an aquarium. They are restless, inquisitive beings

poking their long snouts into every crevice, and
assuming the most extraordinary attitudes. Sometimes four or five of them will be seen quite perpendicular in the water, all having their tails
twisted round the same object, and all holding
their odd little mouths close to the surface, as if
to capture any small insect that might be unfortunate enough to fall into the water. Sometimes
they will assume a perpendicular attitude, but in
just the opposite direction, their tails being near
the surface and their mouths at the bottom of the
aquarium.

These curious positions are maintained by means
of the dorsal fin, *i.e.* that fin which runs along a
portion of the back, and which is in itself a really
wonderful piece of mechanism. Practically, it is
a screw propeller, which undulates instead of revolving, and which causes the fish to advance or
recede in precisely the same manner as the undulations of a snake in the water enable it to swim.
If any of my readers are experienced in boating,
they will understand this movement better by
comparing it with the familiar method of propelling a boat by working backwards and forwards
an oar passed over the stern. Even the rudder
of a boat will act as a screw propeller, if worked
steadily backwards and forwards.

The undulations of the dorsal fin are so rapid
that a somewhat quick eye is needed to discern
them, or even to see the fin at all, which seems to
disappear like magic, and which can only be detected by the reflection of light from the successive
waves that ripple over the fin. The fin itself is
exceedingly thin and delicate, and when the fish is
taken out of the water the fin collapses so completely that it is scarcely recognisabel.

Another of the oddities of the Pipe-fish is the

D

method by which its body is protected.   Instead
of being covered with scales, as is the case with
the generality of fishes, it is armed with a number
of hard flat plates, rather variable in number, according to the particular species.   The reader will
remember that a somewhat similar armature is
found in the Sticklebacks.

But by far the most curious portion of the Pipe-
fish's economy is the pouch in which the eggs are
hatched, and in which the young are sheltered for
some time after they have left the egg.   We are
familiar with a terrestrial example in the kangaroo,
and all the marsupial tribe, and would naturally
expect that certain inhabitants of the waters might
be furnished with a similar apparatus.   In the Pipe-
fish, however, the pouch belongs, not to the female,
but to the male fish, and, in consequence, was the
cause of sore perplexity to those who first investi-
gated it.   The pouch is composed of two long flaps
of skin, which run from the tail along the under
side of the body, and are several inches in length.
Between these flaps the eggs are deposited.   There
they are preserved from the merciless jaws of other
fish, until they are large enough to encounter with
safety the dangers of the seas.

Mr. Yarrell mentions that the fishermen told
him of a curious fact with regard to the Pipe-fish.
If they take a Pipe-fish, open the pouch, and
shake the young into the sea, the little creatures
do not swim away, but hover about the spot, as if
waiting for their parent.   Then, if they hold the
fish in the water, the young will swim to it, and
immediately re-enter the pouch.   Some young that
were found in the pouch of a male fish by the
above-mentioned author measured rather more
than an inch in length.

As to the large tribe of flat fish, such as

Flounders, Dabs, Plaice, &c., they can always be taken in plenty off the shore. Some of them ascend rivers for a considerable distance. The very finest sole I ever saw was caught by myself while trawling in the Medway, just off Gillingham; and I have captured, with a common hand-net, plenty of flounders in the Thames above Erith.

There is a rather amusing mode of catching young flounders. Find a shore where sand is mixed with a little mud, and walk into it with bare feet. The little flat fishes dash about in great terror, and their white bellies glisten as they dart through the water with a speed which few would attribute to them. Presently one of them will be felt wriggling about under the foot. Hold it down firmly, slip the hand cautiously under foot and fish together, and the capture is certain.

No one who has not seen the flat fish swimming can conceive the wonderful grace of their movements. When we look at a plaice or a flounder in a shop window, we think it is about as ungraceful a creature as can well be imagined. And this feeling is not lessened even when we see the living fish in an aquarium, perched, so to speak, on a stone, and with half-raised head keenly watching every object within sight. It has a most uncanny look, and appears as if it were cowering and brooding over some concealed treasure. But when it leaves its perch and begins to swim, it is transformed into one of the most elegant and graceful creatures that can be seen. Instead of merely propelling itself by the movement of the tail, the whole body undulates in a series of curves, the dark brown of the one side and the silvery white of the other contrasting beautifully as the fish writhes its graceful way through the water.

CHAPTER IV.

MOLLUSCS.

THE first of the molluscs which will be mentioned, is that singular creature which is represented in Plate III. fig. 4. This is one of the Squids, a group of molluscs belonging to the family of Cuttle-fishes.

This particular species is common enough on our shores, and may be procured either by taking it in a net, or by hatching it from the egg-clusters which are thrown on the shore at the beginning of summer. The attention of the reader is particularly drawn to these points,—namely, the " siphon," and the eyes. The siphon is the tube by means of which respiration is carried on, and by which the animal ejects the water which has passed into the respiratory system. It is entirely by means of the siphon that the animal projects itself through the water, and this operation is well worthy of observation.

If the water should happen to be very clear, the little creature is seen shooting along without any apparent means of propulsion; but if the water be turbid, or if any fragment of seaweed be suspended in it, the mode of progression is easily seen. When the Squid desires to remain in one spot, the water is suffered to flow gently from the tube ; but if it wishes to propel itself rapidly, it ejects the water with great violence, and so forces itself along in an opposite direction.

Ignorant of this fact, the old naturalists had an idea that the Nautilus, which is a species of Cuttle, used natural sails and oars, whereas the fact is, that the Nautilus simply projects itself by forcing water out of the funnel. Its curious arms are used, not as oars, but as legs, which are quite capable of supporting the body and shell while in the water.

The present species is a lively little creature, and can be kept for some time in an aquarium, where its active habit and inquisitive disposition always attract admiration. The animal is represented of the natural size.

The suckers on the arms ought to be well examined, because they are marvellous examples of natural air-pumps. Each sucker is composed of a fleshy cup, with a sort of piston of the same material. When the edges of the cups are pressed firmly upon any flat substance, and the piston forcibly drawn upwards, the air is extracted, a partial vacuum is formed, and strong adhesion consequently takes place. The "pneumatic" pegs which are now so common, and which are used for hanging objects to shop-windows, flat walls, and the like, are almost exact copies of the sucker of the Cuttle.

The eye is chiefly notable on account of its peculiar structure, which is almost exactly that of the Coddington lens, namely, a transparent sphere with a deep conical groove running round it, and thus producing almost perfect achromatism.

We now come to those interesting molluscs, the Nudibranchs, of which a few are given on Plate III. They derive their name of Nudibranchs, or Naked-gilled molluscs, from the fact that their gills or branchial organs are not placed within the body, but are external and in direct communication with the water. These curious gills are either arranged along the sides or on the back, and in our first examples they are placed along the sides. At fig. 17 is shown a very singular species,—*Doto fragilis*, the latter name being given it because the projecting gills are apt to fall off at a touch. Its general colour is olive or yellow, and its length about an inch. The branchial organs are from six to nine in number, and look something like pine-apples or fir-cones, each being surrounded with rows or projections termed papillæ, which are arranged in regular whorls.

Another species of this genus, *Doto coronata*, is still more beautiful, the yellow of the body being spotted with crimson, and each of the papillæ having a scarlet dot at its tip.

Both of them have the tentacles placed in deep trumpet-like sheaths.

The equally curious creature at fig. 9 is called *Scyllæa pelagica*, the latter title being derived from its oceanic habits. Instead of living at the bottom of the sea, adhering to the growing algæ, or crawling on the rocks, as is the case with most of the Nudibranchs, this species traverses the sea at random, living on floating seaweed, and driven about just where the winds and waves may take it. The tentacles of this species are placed in sheaths like those of the last-mentioned animals; but the branchiæ are borne on four large club-like lobes, which arise from either side of the back. It clings very tightly to the weeds on which it lives, clasping its flattened disc firmly round the stems.

The animal which is called *Eolis papillosa* (Plate III. fig. 13), is a very fine species of British Nudibranch, being sometimes nearly three inches in length. Its general appearance may be seen by reference to the illustration, which exhibits the

mouth, upper surface, and the very numerous
bronchiæ, which are ranged in regular rows along
the sides of its back. This creature is rather
variable in colour, but is generally olive, pinkish,
or brown, though a yellow-specimen is occasionally
seen. It may be found just below the water-mark,
and therefore is to be procured without much difficulty.

At fig. 5, in the same plate, may be seen a
singularly beautiful species, called, on account of
its scarlet colour, *Doris coccinea*. The figure is
enlarged in order to show more perfectly the
circular tuft of gill-plumes in the back, and the
odd-looking tentacles, which can be drawn into
sockets and rendered almost invisible. The gills
plumes are ten in number. The colour of the
species is scarlet, speckled with black. There are
many British species of this genus, all remarkable for the extreme beauty of their colouring,
though in quaintness of form they yield to many
others of the same great group.

In the *Goniodoris nodosa* (Plate III. fig. 8), both
the tentacles and the gill-tufts are much larger than
in the preceding animal, and the tentacles cannot
be retracted. This species is common on our shores,
and may be found without difficulty just below low-
water mark.

Common as it may be, it may easily be passed
over unheeded unless the observer knows exactly
where it is to be looked for; and the same may
be said of all Nudibranchs. If a large bunch of
seaweed be carefully taken out of the water, there
will be found upon it certain little objects looking
like lumps of jelly—brown, olive, pink, and yellowish—but without any determinate shape. These
are certain to be Nudibranchs, and should be carefully removed and placed in water, where they will

soon begin to expand their gill-tufts, and to appear in all their real beauty.

The colour of this species is changeable, being white, yellow, or pinkish, with chalk-white spots. There are thirteen plumes to the gills, arranged in a circle, as in the various species of Doris. Its length is about one inch. We now leave the Nudibranchs and come to another group of the Molluscs.

On all our shores there is a pretty shell popularly known as the Dog Winkle, and technically as *Purpura Lapillus*, which is always an interesting inhabitant of an aquarium, and as it is very hardy, may be kept even by a novice in the management of an aquarium. But any one who keeps this pretty creature must remember that it is as destructive as it is beautiful, and that it will devour almost any other mollusc that it can

Purpura Lapillus.

find. This it does by way of its long, rasp-like tongue-ribbon, with which it scrapes out the unfortunate inhabitant piecemeal after perforating the shell. If it finds a periwinkle or similar mollusc, it does not take the trouble to bore the shell, but passes its tongue into the aperture. But when it feeds upon a mussel or other bivalve, which closes its shell when attacked, it is forced to perforate its shell before it can get at the contents.

Egg-flasks of Purpura.

The eggs of the Purpura are curious little flask-shaped objects of a yellowish-brown colour, which may be seen on the shore in great numbers attached by footstalks either to stones, shells, or even to each other. Each of the flasks contains several young, which are hatched within them, and which remain in their protecting envelopes until they are able to shift for themselves.

This is one of the most variable of shells, sometimes uniformly brown, white, or yellowish; but mostly banded with some other hue.

The name of Purpura is derived from the fact that a deep purple dye can be obtained from a small sac behind the head. The sac is filled with a dark green fluid, which, by exposure to the air, becomes purple-red. The Purpura is supposed to be the mollusc from which the ancients obtained their celebrated purple dye which was only used for the vestments of kings.

As to the common Whelk, which may be found in large quantities along our shores, little need be said of it, except that the aquarium-keeper will be pleased at hatching the young from the egg-clusters which may be found on the shore at the beginning of summer. These clusters are easily recognized, and only need to be placed in

Whelk's Eggs.

sea-water for the young to be produced.

At fig. 6, Plate III., we have an example of the Staircase, or Wentletrapshells, the present species being a very common one. All the species are notable for the bold ridges which ornament them, and for the long spindle-shape of the shells.

Although common on our southern coasts, it is seldom found to the east or northwards; and when it is captured with the animal in a living state, it is well worthy of observation in an aquarium.

The pretty Top-shells are very plentiful along our shores, and can be picked by hand from the brown algæ as they lie uncovered at low water.

Top-shell.

One of our largest, and certainly not the least pretty species, is that which is shown in the accompanying illustration, and which is called *Trochus ziziphinus.* It is rather variable in colouring, but is generally adorned with dots and flame-like markings upon its surface. When the outer coating is removed, an inner pearly coat is seen; and in some foreign species this pearly coat is so beautiful, that the shells are polished and made up into necklaces, bracelets, and other ornaments. When the empty shell is found on the shore, the apex is generally much rubbed, so as to show the shining pearl beneath.

Another species, *Trochus cinerarius,* is shown at fig. 3, Plate III., and is perhaps the most common of all the numerous British species. It may be known by its rather flatter shape, the grey markings, and the narrow perforation through its axis. There is but little gloss on the shell; and in the arrangement of the markings there is considerable variation.

CHAPTER V.

MOLLUSCS AND CRUSTACEA.

THE common Periwinkle (Plate III. fig. 2) can always be procured in plenty on any of our coasts where the brown algæ are found. This mollusc is especially useful if the aquarium should happen to be a glass one, as it feeds upon the young, fresh growth of algæ, which is apt to settle upon the glass and conceal every object behind it. With its long tongue-ribbon it removes the young algæ, proceeding as regularly as a mower with his scythe, and keeps the glass clean without giving any trouble to the owner. The reader is particulaly advised to take out the tongue ribbons of all the molluscs that happen to die, spread them on glass and place them under the microscope, for the purpose of observing the beautiful manner in which the teeth are arranged.

Limpets can always go into an aquarium, and the observer should always place them where they can crawl on glass, so as to show the beautiful mechanism by which they adhere. The same, indeed, is the case with most of the univalves. The common Limpet can be procured in any number; but as it is so familiar, another species has been chosen for the figures. This is the Smooth Limpet, *Patella pellucida* (Plate III. fig. 1). This very beautiful and variable species has the shell almost as transparent as if made of horn and adorned with different colours, such as ultramarine blue, chocolate, yellow, and red. It is to be found among the large Tangle-weeds; and, although not so common as the ordinary species, can generally be found by dragging up large tufts of algæ and examining them carefully.

At fig. 7 is shown an example of the Mail-shells (*Chiton marginatus*), so called because their shells are composed of several plates like ancient mail armour. These plates are not firmly fixed to each other, but have plenty of play in them, and yield to each movement of the animal. Chitons can always be found on the rocks, and their very remarkable structure renders them interesting inhabitants of an aquarium. There are many British species.

The other figures in the illustration represent a few species of the curious beings called Molluscoids, which belong, though with a rather remote relationship, to the Molluscs. Many of them were formerly thought to be seaweeds; while, until a comparatively late period, they were classed among the zoophytes. Fig. 14, Plate III., represents the Sea-mat (*Flustra foliacea*), sometimes called the Lemon-weed, on account of its peculiar odour. Fig. 10 shows the *Gemellaria loricata;* 11, the *Scrupocellaria scruposa;* and 12, the *Bugula avicularia;*—this last being notable as affording admirable specimens of the "bird's head" process when viewed under the microscope. This is, of course, a magnified portion. Fig. 15 represents the *Bicellaria ciliata;* and 16 is a magnified portion of the same. When these are examined in a living state, they are singularly beautiful; each cell being filled with a little inhabitant that spreads its arms in a manner so exactly resembling certain zoophytes, that no one need wonder that the two diverse groups of animals were confounded together.

CRUSTACEA.

We now come to the Crustacea, *i.e.* the Crabs, Lobsters, Shrimps, and their kin, many of which can be kept for a considerable time in an aquarium.

The most common of all our seaside Crustacea is the familiar Green Crab, *Carcinus mœnas* (Plate IV. fig. 1), which may be taken in almost any number, either by the hand, the net, or the line.

This Crab is not considered as eatable, though, in fact, it is nearly as good as the true edible Crab of our shores; but the fact is that the meat is so small in proportion to the shell, that it hardly repays the trouble of catching and cooking.

As its name implies, the predominant colour of this crab is green, though it is extremely variable, and is seen of various tints. Sometimes it is quite a deep brown; sometimes the whole of the under surface is orange; while specimens are not at all uncommon in which the hue is yellow, with a black mark on the back. These are generally the small young specimens, while the darker hues belong chiefly to the largest and oldest specimens, some of which are brown, with just enough tinge of green to justify the popular name.

These latter individuals are not at all adapted for the aquarium. They require much more space than can be afforded to them, and besides are so voracious that they would eat every other inhabitant, not excluding those of their own species. In fact, there is nothing that a large crab seems to like half so much as a little crab; and even on the shore, crabs may be frequently seen pursuing one another with cannibalistic intentions, the little one only escaping by creeping into some hole which its larger relative cannot enter.

They are very hardy animals, and can be conveyed for considerable distances with perfect security. Should the reader be disposed to bring crabs from the seashore, he will find that the best mode of conveying them is simply to put them into a can, with a quantity of wet seaweed of any of the green varieties. If brought in water, they are apt to die; but if merely packed up in wet seaweed, they may be kept for several days without touching the water.

Continued immersion in the water of an aquarium is not good for a Green Crab, and the creature will be in better health if a few stones be piled up in the middle, so as to allow the crab to get out of the water without being able to escape. Even being occasionally removed from the water, and permitted to run about on the floor of the room, will be useful to this amphibious and active being.

The voracity of the Green Crabs is almost incredible, and their agility little less so. As to their performances in the water, little is known of them, except that they annoy anglers exceedingly by taking their baits, deluding them into the idea that a fish is on the hook, and obliging them to rebait the hook. But on land I have often watched them, and been much interested by their proceedings.

They can best be seen when the tide is flowing; and if any one will take the trouble to remain perfectly motionless, he will see one method by which the Crabs obtain food. They are hardly recognizable, full of life and fire, darting here and there as they suspect the presence of prey, pursuing everything that seems to have life, anchoring themselves in the sand by their sharp feet, and inspecting every object as the tide passes by them. Even the agile Sandhoppers are chased and caught:

by the Crabs; so are the Sandflies; and I have seen them pounce on bees that came to drink the salt water, encage them within their legs, pick them out daintily with their claws, hold them in one claw, pull them to pieces with the other, and eat them.

Therefore, if a Green Crab is to be kept, give him a house to himself, feed him liberally, and do not afford him the least chance of escape. A crab can climb wonderfully well, and our Green Crabs, although they cannot rival the great Ou-Ou Crabs of Samoa, which climb the cocoa-nut palms and throw down the fruit, will make their way out of vessels which were thought to be quite safe. They have a fashion of getting against the side of the vessel, scrambling up it as high as they can, hitching the feet by one side over the edge, and then pulling themselves up. If a crab succeeds in putting the tip of a single leg over the top of the vessel, it is sure to make its escape.

Should the reader like to catch some Green Crabs for himself, he can handle the oldest and fiercest of them without danger of being caught by their sharp nippers. These crabs are capital fighters, quick as a modern boxer in delivering a stroke, and grasping so forcibly with their sharply pointed claws that they occasion no little pain. But there is an infallible mode of capturing the creature without danger of a bite. When a crab is brought to bay, it sits upright, holding its claws extended so as to guard the whole of its body. Make a feint at it with the left hand, at which it will instantly strike with both claws, snapping them audibly as it does so. At the same moment, push it forward with the forefinger of the right hand, and press it firmly against the ground, while with the thumb and middle finger, the body is held tightly, just behind the claws. It is better to press the claws slightly towards each other, and when in this position the Crab is completely disarmed.

Very soft specimens may be often found on the shore, mostly hiding under rocks and stones. They are perfectly helpless, and so feeble that if they are picked up hastily, one or two of their limbs will probably fall off. These are crabs which have just cast their shells, and which have not as yet been able to secrete a new one fitted for their enlarged bodies. They are obliged to hide themselves, or they would certainly be eaten by their hard-shelled relatives. On some of our coasts these soft crabs are called "Peelarts," and are used as bait.

At Plate IV. fig. 2, may be seen a figure of another species, the Harper Crab or Sea Toad (*Hyas araneus*). As its latter name implies, this creature belongs to the group called Spider Crabs. All these crabs may be known by the shape of the body, which has a pointed snout. Owing to the shape of the body and the manner in which the limbs are set upon it, the Spider Crabs are not so active as the Green Crabs; and, indeed, often have their bodies so thickly covered with zoophytes that the shell can scarcely be seen. Another allied species, the Thornback Crab (*Maia squinado*), is almost invariably covered with these parasites.

The Harper Crab is held in great detestation by those salmon-fishers who use fixed engines. When the tide happens to ebb during the night, and the salmon are left high and dry, these crabs are sure to come and revel upon the dead fish, injuring their appearance and destroying much of their saleable value. It may generally be found about low-water mark, and is not so fond of scuttling over the sands as the Green Crab.

One species of the Spider Crab, namely, the Slender-legged Crab (*Stenorhynchus phalangium*), is tolerably common on our coasts, and is found mostly where the bed of the sea is formed of sand and mud. A figure of this species is given at Plate IV., fig. 5. Mr. Hillier tells me that he has noticed a very curious habit on the part of this crab, specimens of which he has kept in his aquarium. After the crab had shed its shell, it was seen to take a quantity of seaweed and place it over its body, as if to guard itself in its soft and helpless condition. Perhaps this fact will account for the quantities of algæ that are often seen growing upon the backs of this and allied species.

If the reader will look at the legs of the Green Crab, he will see that the last pair are flatter than the others, especially towards the last joint. This structure shows that the creature belongs to the Swimming Crabs, although in that species the feet are not sufficiently flattened to allow it to swim. Some exotic Swimming Crabs have these limbs as wide and flat as the paddles of a canoe, and by this means the creatures are propelled through the water as swiftly as the fishes themselves.

We have several species in England in which the paddle-like feet are tolerably developed, although the limbs are not so wide or so flat as in the tropical species, and consequently their owners cannot swim nearly so fast. One of the best known of the British Swimming Crabs is the Velvet Fiddler (*Portunus puber*), which may be easily recognized by the accompanying figure.

It is called a "Fiddler" because in the act of swimming it works its hinder legs backwards and forwards with a quick movement, something like that of a Fiddler's arm. All the Swimming Crabs possess this movement, and are popularly called

E

by the same title. The name of Velvet Fiddler is given to it because its body is covered with a thick coating of soft velvet-like hair. This, al-

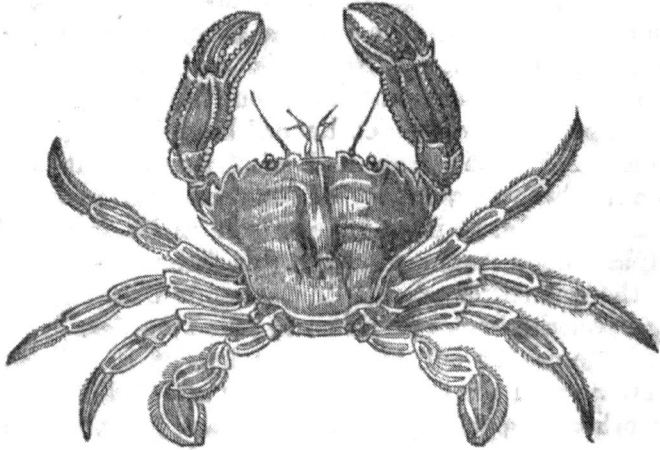

Velvet Fiddler.

though not a large species, is a very handsome one, the claws and legs being striped with blue and scarlet, which contrast beautifully with the rich brown hue of the velvet-clad back.

Like the Green Crab, this species is eatable, though not generally recognized as such, and in some places is considered as a great delicacy. It is sold for the table under the name of Lady Crab. These little crabs are apt to get into the lobster-pots in search of the bait which is placed there to attract more valuable crustacea, and, as a general rule, they are flung away as worthless.

In an aquarium, they are quite as interesting as the Green Crab, though not so gene-rally active, nor so obtrusively voracious. But although they greatly prefer quietude, keeping themselves closely under shelter, they are ready

enough to come out when they see anything that
looks eatable, and to do their best to eat it, whether
it be alive or dead. Smaller crabs of any species
are sure to fall victims to the Velvet Fiddler, while
flesh of any kind is always acceptable to these voracious crustacea.

We now come to a series of crabs which are
remarkable for their curious development. Instead
of being covered with a hard shelly coat of mail,
only the fore-parts are so defended; and the abdomen, which is long and projecting, is as soft as
if the animal had only just shed its shell. In
order to protect this soft abdomen, the crab inserts
it into the shell of some univalve mollusc, and
carries the shell about just as if it were the lawful
owner of the usurped house.

These animals are known by several names. On
account of their solitary habits, each living in a

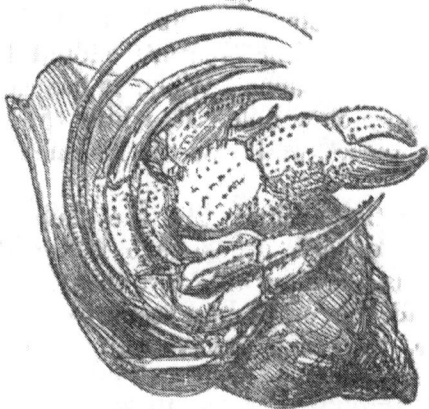

Soldier, or Hermit, Crab.

separate shell, they are sometimes called Hermit
Crabs, while their singular aptitude for fighting
has gained them the name of Soldier Crabs. The

most common of these is the ordinary SOLDIER
CRAB (*Pagurus Bernhardus*), which is represented
in the accompanying illustration as it appears
when inhabiting the shell of a whelk. The same
animal is also shown at Plate IV. fig. 8, as it
appears when removed from the protecting shell.

The reader will note that at the end of the
abdomen there is a kind of double hook. This is
the clasper by which the crab maintains its hold
of the shell; and so firmly does it grasp the shell,
that it may be pulled to pieces before it will loosen
its hold. The entomological reader will remember
that the caddis-worm is furnished with a similar
apparatus.

The claws are unequal in size, one being very
much larger than the other. When the crab
withdraws itself into its shell, the large claw
effectually closes the entrance, and is doubly effective, first as a door, and then as a weapon.
With this large claw it seizes prey, as will soon
be seen if any living creature should happen to
come within its reach. When it has captured its
prey, the large claw holds it, while the little claw
picks it to pieces, and puts the fragments into the
mouth.

Although the animal and the shell are mostly
well suited to each other, such is not always the
case, and it is a remarkable fact that, however well
the shell and the crab may seem to be suited to
each other, the crab always thinks that a shell
belonging to another crab would make a better
house. Consequently they will wage direful battles
over a few empty shells, although neither of the
shells would make so commodious a habitation as
that which was already occupied.

If the reader is disposed to find a specimen or

two of the Porcelain Crabs, he can do so without
much trouble. He should choose the time of extreme ebb tide, wade into the sea just about low-
water mark, and turn over all the loose stones that
he can find. Under these stones may often be
found the BROAD-CLAW CRAB (*Porcellana platy-*

Broad-claw

cheles), a creature which might easily be unnoticed
because of its flat body, and the mode in which
the limbs are packed so closely to the body that it
looks at first more like a stone than a crab.

There are many species of this group, and their

popular name is given to them on account of the
shining porcelain-like surface of the under side of
the body. The name of Broad-claw is given to
this species because its claws are very broad and
very flat, so as to allow their owner to creep into
a small crevice. They are also furnished with a
beautiful brush-like apparatus, by means of which
the animal sweeps the sea, and carries towards its
mouth the minute particles of edible matter that
would otherwise be wasted.

An apparatus of this kind is an absolute necessity for the Broad-claw Crab, because it does not
wander about in search for food, but remains
quiescent in one spot. It is therefore gifted with
a peculiar apparatus, by means of which it can
obtain subsistence as the water flows past its place
of residence.

WE now come to the Long-tailed Crustaceans,
of which the Lobsters, Prawns, and Shrimps are
familiar examples. These creatures have a vast
amount of muscular power thrown into their
"tails," by means of which they can project
themselves through the water with astonishing
velocity. If the fan-like apparatus at the end of
the body be examined, it will be found to consist
of several shell plates, so arranged as to offer the
greatest possible resistance to the water when
drawn towards the head, and as little as possible
in the other direction. So when a lobster wishes
to dart off in a hurry, it simply contracts its body
suddenly, and so drives itself backwards through
the water with such a velocity that the eye can
scarcely follow its course. Any one who has
watched the shrimps and prawns disporting themselves in the shore-pools will be familiar with the

quick, darting movements of the little creatures.
They crawl forwards by means of their legs, but
they dart backwards by means of their tails.

Two very familiar examples of these crustaceans
have been selected, both of which are easily kept
in an aquarium. The first is the common SHRIMP
(*Crangon vulgaris*), which may be obtained in
multitudes on any sandy shore.

The Shrimps may be easily caught either by
hand or by net, the latter being the preferable
mode, as they run less chance of being injured.
The best method of obtaining good specimens, in
perfect condition, is to waylay a shrimper and give
him a few pence to allow you to make your choice
of the contents of his net, which is a much more
effective article than the ordinary hand-net A
shrimp of the usual size is shown at Plate IV.
fig. 9.

At fig. 7 of the same plate is a figure of one of
our British Prawns, popularly called the White
Shrimp (*Palæmon squilla*).

The Prawns are always among the most beautiful
inhabitants of an aquarium, their transparent bodies

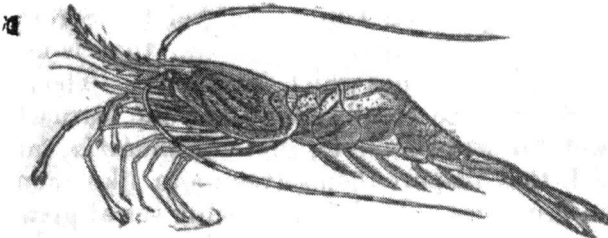

Æsop Prawn.

painted with different colours, and their eyes glowing as if they were balls of phosphorus. The effect
of a Prawn's eyes in an aquarium is often very

remarkable, the only part of the creature that is visible being two little globes of fire moving steadily through the water. There are many species of British Prawns, among which the Æsop Prawn (*Pandalus annulicornis*) is among the most beautiful. This lovely little crustacean is called the Æsop Prawn because it has a hump on its back, and the name of *annulicornis*, or ring-horned, is given to it because its antennæ, or feelers, are ringed with scarlet at regular intervals. Its body is transparent grey, slightly tinged with green, and varied with scarlet lines.

Figs. 3 and 4 are examples of those crustacea which are called Sessile-eyed, because their eyes are placed directly on the head, and not set on footstalks. The first of these is the curious little creature called *Corophium longicorne*. It inhabits muddy shores, and lives in little holes, with which the mud is often perforated like a honeycomb. It is a voracious little being, and will eat marine worms, molluscs, and even the fry of fish, if they come in its way. By way of poetical justice, they are much eaten by gulls and other birds that come to feed at the edge of the sea.

Fig. 4 represents Pennant's Skeleton Screw (*Caprella linearis*), a little crustacean that is notable for the odd attitudes which it assumes. The Skeleton Screws are chiefly found on the barnacles of a well-known zoophyte, the *Plumularia*, among which they disport themselves more like monkeys than crustacea. Their limbs being set at great distances from each other upon their slender bodies, causes their movements to bear some resemblance to those of the "looper" caterpillars, although they are much quicker and more varied.

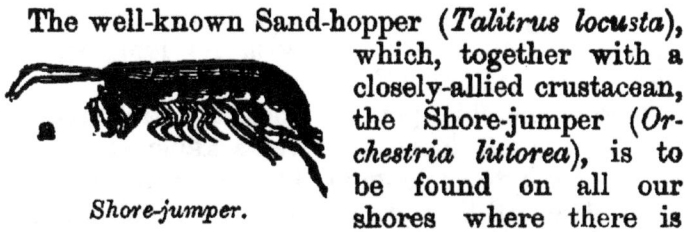

The well-known Sand-hopper (*Talitrus locusta*), which, together with a closely-allied crustacean, the Shore-jumper (*Orchestria littorea*), is to be found on all our shores where there is

Shore-jumper.

the least modicum of sand. Both these creatures are so similar in their forms that they may easily be mistaken for each other, while in habits they are identical. The chief distinction between them is that the latter species has the joints of the two first pairs of legs modified into claws, which is not the case with the Sandhopper.

Both these creatures are to be seen skipping and hopping about the sand in great numbers, especially as the tide rises, when they emerge by myriads out of the sand into which they have burrowed. But if the wanderer by the seashore wishes to see them in greatest numbers, he should choose a time when the tide is ebbing, and walk along the shore about half tide. The heaps of seaweed which are left by the retiring waters, and especially if there are corallines among them, literally swarm with Sand-hoppers, which are under them, and in them, and round them, and piled on them in heaps, which rapidly dissolve with a hissing as the spectator approaches. Sometimes they are so numerous, and leap so perseveringly, that the continuous fall of their bodies on the sand resembles the pattering of a hail-shower.

The last crustacean which we shall mention is the strange being called *Phoxichilidium coccineum*. It has no popular name. I took one of these creatures at Margate in a tuft of green alga, and kept it for some time, both at the seaside and in my own house. It was a strange-looking creature, never doing anything in particular, but continually sprawling about fatuously with its long, thread-like limbs. (Plate IV. fig. 6.)

One of the chief points of notice in this creature is that it is developed in a remarkable appendage to a zoophyte belonging to the genus Coryne. In a little ball-like growth, the Phoxichilidium is packed away, its long limbs being rolled round its body just as a ball of twine is wound up. It is not very common, but may be found by pulling up large tufts of seaweed, carrying them home, and examining them carefully in sea-water.

CHAPTER VI.

ANNELIDS.

WE now come to a series of creatures which apparently are not so interesting as those which we have already examined, because they do not possess the same amount of apparent life, neither are their movements so easily understood. But they only need to be watched carefully, and they will then prove to be quite as interesting as the animals already described.

We will begin with those which belong to the great class of Annulata, or the Worm tribe, and take as our first example the SEA MOUSE (*Aphrodite aculeata*), Plate V. fig. 7. I need hardly remind my classical reader that Aphrodite was the Greek goddess of beauty, who sprung from the foam of the sea. The name has been given to the animals belonging to this genus of marine worms on account of their marvellous beauty. Most of the worms are furnished with hairs or bristles, generally very short, and not perceived except by careful eyes. The Aphrodite, however, has the sides of its body so thickly covered with hairs and bristles, that they overshadow its back and hide it from view.

The beauty of these hairs is almost inconceivable. They are as magnificently coloured as the plumes that decorate the throat of the Humming Bird, and, unlike their feathers, change their colour with every movement. Every colour of the rainbow is reflected from these hairs; and as each individual hair reflects a different colour, it may be imagined that the appearance of the animal is indeed beautiful. Why the creature should be endowed with such a gorgeous dress is not easy to see, because it lives where the light seldom penetrates, and where its beautiful clothing is all hidden. The chosen habitation of the Aphrodite is in the muddy bed of the sea, and, as if not satisfied with hiding its beauties in the black and fœtid mud, it creeps beneath stones or shells, so as to be completely hidden from the light.

Whatever may be its habits in the sea can only be conjectured from its dimensions in an aquarium, where it generally lies in the same spot, and appears to take no delight in moving. Herein it displays a great contrast to the generality of creatures which are distinguished for their beauty, and which seem to take the greatest delight in exhibiting their perfections to the best advantage.

Sluggish as it is, the Aphrodite is a dangerous creature for the aquarium, as it is exceedingly voracious, and will eat almost any other inhabitant that is not defended by a shell or hard armour. It does not even spare its own kind, and has been known to eat a companion in two days, in spite of the struggles of its unfortunate victim. The instrument which it uses in this predacious manner is a proboscis, which can be thrust out to a considerable distance.

Should any of my readers procure a living Aphrodite, which may easily be done by means of the dredge, they are strongly advised to examine the structure of the breathing apparatus, which is almost if not unique in the animal kingdom. By pushing aside the hairs that cover the back, a number of flaps will be seen, which rise and fall very regularly. These are, in fact, portions of a set of forcing-pumps, which admit water when they rise, and drive it through the respiratory organs, where they unite. The hairs which cover the valves act as filters by which the water is strained clear of the mud in which the animal lives, and therefore the felt-like covering which is

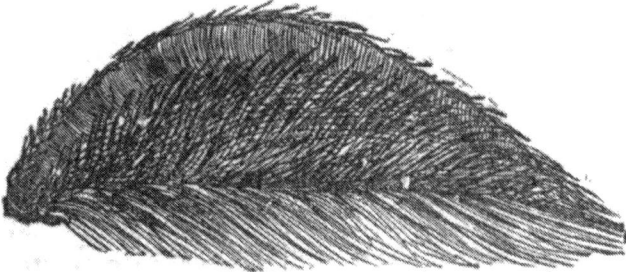

The Sea Mouse.

formed by them is always so full of mud that it cannot be cleaned without great difficulty. Indeed, a specimen which is to be dissected requires frequent washings and rinsings before it can be placed in the spirits of wine in which it is dissected.

Some species which live in places where the water is clear do not need the hairy covering of the valves, and therefore do not possess it.

In order to show the appearance of the Aphrodite when crawling, a side view of it is given in the accompanying illustration.

On several parts of our coasts, more especially on the southern shores of England, the wanderer by the seaside may find great numbers of a very interesting marine worm, which lives in tubes made out of sand. On account of the material from which they make their habitation, the creatures belonging to this group are called by the name of Sabella, and the best-known species is that which is known to naturalists as *Sabella alveolaria*. This worm is an eminently social being,

and always builds its tubes in agglomerated masses.
A portion of one of these masses is shown in the
illustration, as it appears when cleaned from the
loose sand that lies among the tubes, and almost
renders their form invisible.   The process of

Sabella.

clearing away the sand is not a very easy one,
as the tubes are exceedingly fragile; and crumble
to pieces if handled roughly.

The strength of the tube is, however, variable,
and when the tubes are attached throughout their
length to a piece of rock or stone, they are very
much stronger than when they are comparatively
free. A specimen now before me is of the latter
character, and can be crumbled to pieces between
the fingers.

At *a*, in the illustration, one of the inmates may
be seen protruding from its tubular home, and at
fig. 6, Plate V., a specimen is shown as it appears
when removed entirely from the tube and magnified
to twice its usual dimensions. When in its tube,
the slender tail is doubled up along the body. The
beautiful tentacles which adorn the head are most
elaborately constructed. They are about sixty in
number, and each is furnished with a double row

of teeth, very much resembling the arrangement of the teeth in a double comb. In order to enable the animal to emerge from its dwelling and' to withdraw itself at will, many of the segments are furnished with small but strong bristles. As is the case with most of the annelids which inhabit tubes, the creature protrudes its body very slowly, but withdraws it with so quick a jerk that it seems to vanish by magic.

The work of tube-building goes on almost without interruption, though the Sabella appears to prefer the night as its time of labour. Sometimes the animal leaves its tube altogether, and appears to be as much at ease out of its residence as in it. Many of the tube inhabitants follow the same custom, but, as a general rule, they only come out to die.

If a number of Sabellæ are placed in a glass vessel, and supplied with plenty of sand, they will build their tubular houses as freely as if they were in the open sea, especially if the vessel be kept in darkness. When they can be induced to do so, the aquarium-keeper may rejoice, as he will have an excellent opportunity of watching the habits of the animal. Like many other creatures that construct houses, and are supposed to act from instinct alone, the Sabella possesses a sufficient amount of reasoning power not to work more than is absolutely necessary. If it finds itself placed near the side of a glass vessel, it will attach its tube to the glass throughout its entire length, and will not only gain a permanent support for its habitation, but will make its tube a mere segment of' a cylinder, the glass supplying the missing portion. Thus the observer can see the tube and watch the proceedings of its inmates without the least difficulty.

Our next example of the tube-inhabiting worms is the beautiful Serpula (Plate V. fig. 9).

This creature derives its name of Serpula, a little serpent, from the snake-like coils of its tube. Its tube, instead of being made of sand, like that of the Sabella, is formed of a calcareous or chalky substance, and is as white and nearly as hard as white china.

In general structure the worm which makes and inhabits this tube resembles the Sabella, but it is far more beautiful. The double fan of tentacles which project from the head is of a brilliant scarlet, variegated with white, and one of the tentacles is modified with a conical stopper, deeply grooved, which enables the Serpula to close at will the entrance of its tube. Sometimes, especially in captivity, the Serpula remains obstinately hidden in its tube, the lovely fan concealed, and the stopper jealously guarding the entrance.

And so sensitive is this little worm, that even when it is fully expanded, a hasty footstep in the room, a sudden jar to the floor, or even a bird flying between the light and the aquarium, will cause the Serpula to start back with such rapidity that the eye cannot follow its movements. This last fact is very singular, especially as no vestige of visual organs can be seen. But that the animal is exceedingly sensitive to light and darkness is a fact, for which any one who has kept Serpulæ can at once vouch. I have had many of these lovely creatures, and have always admired the extraordinary sensitiveness possessed by a being which looks as if it had no senses at all. Like the Sabella, the Serpula occasionally leaves its tube ; but I have always found that such a proceeding means a speedy death. For my own part, I was always rather glad when they did leave their tubes to die,

as they could then be easily removed from the
aquarium, and did not taint the water with their
dead bodies.

This particular species, *Serpula contortuplicata*,
can seldom be obtained without the aid of the
dredge; but there are two species that can be
procured at low water without the aid of any
instrument except the hand. One is the *Serpula
triquetra*, so called because its tube, instead of
being round, is triangular, like a bayonet; the
three corners being as well defined as in that
formidable weapon. The tube of this species does
not stand out freely like that of the preceding
Serpula, but is attached throughout its whole length
to some object, generally a stone or an oyster shell.

The last species which will be mentioned is a
very small one, that is found chiefly on the stalks
and fronds of the great Oar-weed, or Tangle (*Laminaria*). If the young observer will go to the seashore at low water, and wade boldly into that
portion of the shore which is technically named
the Laminarian zone—*i.e.*, that part in which the
Laminaria grows—he will see that the fronds and
stalks are beset with tiny spinal shells set flatly in
the leaf, and looking very much as if they were
univalve Molluscs in a very early stage of existence. They are, however, the homes of veritable
Annelids, which are known to zoologists as belonging to the genus Spirorbis. These worms closely
resemble the true Serpulæ, and, in spite of their
minute size, possess a stopper almost exactly the
same as has been already described.

It is not at all necessary to go as far as the
Laminarian zone in order to find specimens of the
Spirorbis, as they are to be seen on almost any
piece of seaweed that is large enough to afford
them a resting-place. But if they are needed for

the aquarium, it is better to go as far seaward as possible.

One of the commonest and, at the same time, one of the most interesting of tube-building worms, is that animal which is popularly known by the name of Sand Mason, or Shell-binder, according to the materials of which its habitation is made, and is scientifically termed *Terebella.* One of these worms, the Shell-binder, is shown at Plate V., fig. 8, as it appears when partially pro-truded from the tube. These creatures are in fact two distinct species, but the structure of them-selves and their habitation is so similar that they need not be de-scribed separately.

This worm may be found on many of our sandy shores, and may be easily distinguished by the thread-like tentacles that proceed from the mouth of the tube. These tubes generally protrude about an inch or so from the sand, and their appearance may be learned by re-ference to the accompanying illus-tration. The tubes are about a foot in length, and have two openings, one at some little distance from the other; and the inhabitant is able to move at will to either end of its tube, aided by the tufts of bristle with which some of its segments are furnished.

Terebella.

Unlike the fragile tube of the Sabella, or the hard and brittle tube of the Serpula, that of the

Terebella is tolerably tough; the fragments of which it is composed being bound together by a secretion from the animal which has a silken aspect, and not only makes the materials of the tube adhere to each other, but forms a smooth and soft lining. If a perfect Terebella and tube be wanted, the only way is to dig very cautiously under the sand, the touch being exercised more than the sight. As in rocky places the worm has a habit of burrowing under a piece of rock, and leaving an end of the tube at each side of it, the perfect tube is not to be obtained without much care and trouble.

But when it is dug out, the aquarium-keeper has a treasure which will repay him for any amount of labour. If he likes, he can turn the worm out of its tube, and by supplying it liberally with sand or broken shell, according to the species, can see it build another tube around itself. This process is achieved by means of the tentacles, which form themselves into tools of marvellous power. Interesting as is the mode in which the tube is built up, it is so complicated that our space will not allow it to be described, and in consequence the reader is advised to catch the worm and watch it for himself.

THE RADIATED ANIMALS.

Our next group of marine animals is called by the name of Radiata, because their parts radiate from a common centre. They comprise the Star-fishes, the Sea Urchins, the Sea Cucumbers, and their kin, and we will examine briefly one or two examples of each.

The first example is the Sea Cucumber (*Holuthuria*), which is represented in the accompanying illustration. It is generally obtained by the dredge, as it does not frequent shallow water, on account of the dislike which it bears towards light, and has a habit of clinging to the under surface of stones and similar objects, so that stones and animals are hauled up together. One of the most curious points in the Holuthuria is the number of slender white tubes with which the body is almost entirely filled, and a habit among some species of committing suicide by beating themselves into several pieces. Sometimes it will throw off its beautiful feathery coronet of tentacles, as well as the whole of its digestive apparatus. However, if allowed to remain in tranquillity, it reproduces the whole apparatus, and appears in perfect health. Many species of Holuthuria are eaten by the Chinese under the name of Trepang.

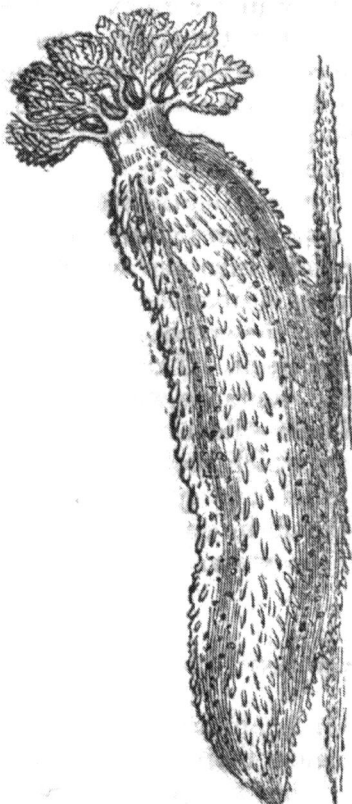

We now come to the Sea Urchins, or Sea Hedgehogs, so called because they are covered with spines that bear some resemblance to the quills with which the hedgehog is armed. The spines are moveable, each being furnished with a ball-shaped base, which works in a socket attached to the body of the animal. The attachment of the spines is exceed-

Sea Cucumber.

ingly slight, and when the creature is dead, they fall off with the least touch. When these spines are removed, the sockets can easily be examined, and have a very pretty appearance, being ranged in regular order upon the outer case of the animal.

This outer case, or shell, is well worthy of notice. It is composed of hexagonal plates of shelly matter, and as each plate is continually enlarged by the deposition of fresh sub-stance round its edge, it will be seen that the shell can grow con-tinually without alteration of its shape. There are about ten thousand of these plates in a single specimen. In the accompanying illustration may be seen a portion of the shell and spines. On the right are two of the spines removed from their places, so as to show their globular bases, while a number of the sockets are seen on the plates occupying the shell. A portion of four plates is given, to show the mode of junction, and the manner in which they can be regularly increased in size without altering in shape.

This part of the structure of the animal must, of course, be examined after its death. During its life, the Echinus does not seem to be particu-larly active, though it will walk up and down the stones within the aquarium, or even up the glass sides. If it can be induced to perform this latter feat, the observer should carefully notice the won-derful means of progression which it employs. These consist of a vast number of organ-like tentacles, which project through minute holes in the case, or box, of the Echinus, and which answer the purpose of feet. In some species there are at least fifteen hundred of these feet.

To watch the creature in the act of walking is

most interesting. Innumerable feet project from the shell, and attach themselves to the glass, some bending in one way and some in another, but all with a definite object. Sometimes the animal twists itself quite round by means of these curious organs, and it is most extraordinary that their action should ever be sufficiently under control to enable the animal to direct its course. An Echinus is seen at Plate V., fig. 1.

The structure of the mouth is also worthy of notice. If the observer will look at the orifice at the lower surface of the Echinus, he will see that five very sharp tooth-like objects meet in its centre. These are the tips of the teeth, and are the only visible portions of a most wonderful and complicated structure. Our space will not allow this marvellous organ of mastication to be described, and, indeed, the object of this little work is chiefly to direct the owner of an aquarium to the points most worthy of investigation, and not to describe them. Suffice it to say, that if a portion of the side of the shell be carefully cut away, the whole formation of the mouth, with its five long, sharply-pointed teeth, and the lantern-like form of their arrangement, can be easily seen. The examination should be made while the creature is still fresh, so that the teeth can be moved and the action of the muscles seen.

If the observer will take an ordinary magnifying-glass and examine the surface of the shell, he will see that between the spines it is studded with a vast number of tiny projections called pedicellariæ. They can be seen easily enough without the aid of the glass ; but their form is better made out by the use of a lens. Their stems are hardly thicker than hairs, and their tips are furnished with round knobs, or, in many instances, with three bag-like blades, which are continually being opened and closed. During the life of the Echinus, the pedicellariæ are in constant motion, swaying about from side to side, and the strangely-formed tips opening and shutting their blades with singular regularity.

CHAPTER VII.

THE STAR-FISH.

THE accompanying illustration represents a very familiar example of the next group of radiates, namely, the Common Five-finger Star-fish of our coasts, which may be found in such vast numbers along the shores, especially after a gale. Its scientific name is *Uraster rubens*. Common as is

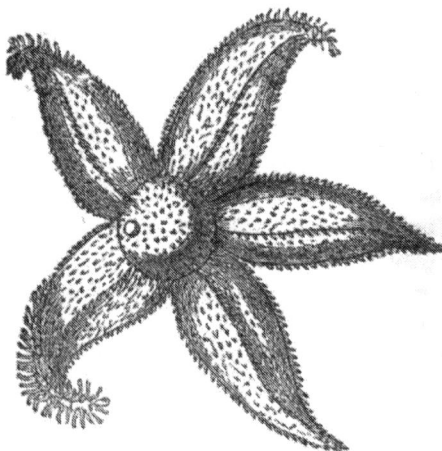

Five-finger Star-fish.

this creature, and much as it is despised by the generality of persons, it is well worthy of a very close investigation. Let one of these Star-fishes be taken up and its under surface examined. If it be alive, thousands of "feet" similar to those mentioned when treating of the Echinus will be seen protruding from the under surface of the rays, moving about and twisting their translucent stems in every direction, as if trying to find some object on which to gain a hold.

By means of these "ambulacra," as they are scientifically termed, the Star-fish contrives to proceed at a regular though slow pace, and has a marvellous facility of gliding through passages which seem far too narrow to permit the body to pass. Even if it be laid on its back, the Star-fish can turn itself over by means of these ambulacra, bending the rays so as to fix as many of them as possible, and then gradually drawing itself over by the attachment of more and more of these organs. In the illustration, the lower limb on the left-hand side is slightly turned over, so as to show the general appearance of the ambulacra. The working of these beautiful organs of progression is best seen by taking the animal home and putting it in the aquarium. Experiments can be tried by surrounding it with stones, and seeing how it either surmounts the obstacles or creeps between them. In either case it accommodates itself to the shape of surrounding objects in a very curious manner.

The aquarium-keeper must not put the Five-finger in the same vessel with other animals, especially with molluscs, as it is a most voracious being, and will do almost as much harm as a wolf in a sheepfold. It eats them by clasping its five limbs round them and forcing them into its mouth, if they are small; and if they are too large to be swallowed, it contrives to eat them out of their shells. Oyster-beds suffer greatly from the attacks of the Five-fingers, and even when an angler is fishing in the sea he is apt to find that his bait has been taken, not by a whiting or other fish, but by a Five-finger. Crabs are annoying enough to the marine angler, but the Five-finger is scarcely less troublesome, wrapping its red rays round the bait, and effectually keeping off the fish which it was intended to allure.

In consequence of its predatory habits, professional fishermen have a deadly hatred of this species, and destroy, or rather try to destroy, every specimen that they capture. But, owing to their ignorance of zoology, they fail most signally in their endeavours. Like many other beings that occupy a very low position in the animal kingdom, the Star-fishes have a very great power of reproducing lost or damaged members. It is a common thing to find specimens of this very species with only three rays, and sometimes a Five-finger is found with only one ray complete. Now, the fishermen, either ignorant of their reproductive powers, or failing to draw a just inference from their knowledge, have a custom of tearing the Star-fish in two, and throwing the pieces back into the sea. The consequence is, that the two halves become two individuals, and the fisherman has only doubled the power of his foe instead of destroying it.

With care, and, above all, with a plentiful supply of frequently-renewed water, a very cool locality, and a judiciously darkened aquarium, these creatures will live for a considerable time, and will allow their habits to be studied. But, like all inhabitants of an aquarium, they are sure to die sooner or later ; and when the event takes place, their structure ought to be examined. There are three points to which I would specially draw attention. The first is the structure of the hard, rough skin which covers the upper surface, together with the pedicellariæ which are found upon it.

The next point is the mouth, and the curious arrangement of the digestive organs ; and the third is the elaborate skeleton, if we may use the word, by which the form of the creature is preserved.

If the rays of a common Star-fish be merely laid open, a most wonderful structure presents itself, rows upon rows of pure white pillars standing in regular order like fairy colonnades. The whole of this beautiful structure can be laid bare by washing the Star-fish well in fresh water, and then putting it near an ants' nest. The Star-fish should be in a box pierced full of holes, so as to allow the ants to enter and leave the box, while the dust and wind are kept out.

At Plate V., fig. 3, is a ·representation of another common species, which certainly has a better claim to the title of Star-fish than that which has just been described, the rays being twelve in number instead of five, and ranged regularly round the central disc. This is the common SUN STAR (*Solaster papposa*). This fine species occasionally attains a considerable size, and has a very handsome appearance.

At Plate V., fig. 2, is shown one of the remarkable creatures called Brittle Stars, because they have a habit of breaking themselves to pieces whenever they are alarmed. It seems almost strange that such a creature should experience a feeling of alarm, or, indeed, any mental emotion whatever. Yet the Brittle Stars are peculiarly timid, and have some strange instinctive way of detecting danger. One would think that, however the danger might be dreaded, it could do no worse than beat the Star-fish to pieces, and yet the creature adopts this singular mode of escaping from its enemies.

The species which is represented in the engraving is one of the commonest of the English Brittle

Stars, and can be found on almost any of our coasts.
For large specimens a dredge is needful, but for
very small examples no tools are required, except
perhaps a little pail.

If the young naturalist will go down to the
shore at low-water, and wade into the pools and
runnels worn in the rocks by the action of the sea,
he will perceive masses of green algæ, of which the
common Cladophora is best for his purpose. Let him
grasp the bunches of Cladophora just above their
attachment to the rock, pluck them off, and put them
into his pail, and take them home. He should
then spread out separately each bunch of seaweed in
a basin of sea-water, and in almost every bunch he
will find some Brittle Stars of very small size, but
quite perfect. One or two of the best of these
should be preserved as objects for the microscope,
and, if possible, should be examined with a rather
low power—say an inch focus—and with a bino-
cular instrument.

There are few inhabitants of the aquarium which
are more familiar than the Actiniæ—Sea Anemones, as they are popularly called. These singular
examples of animal life are plentiful in almost every
part of the world, although they do not attain their
full luxuriance of colour in the colder seas, and in
the Polar regions appear to be absent altogether.

On the British coasts we find a wonderful variety
of these animals, the largest and finest specimens
belonging almost exclusively to the southern shores,
and being especially plentiful in Devonshire. The
most common of the British Actiniæ is shown in
Plate VI., fig. 1, as it appears when expanded.
Scientifically it is called *Actinia mesembryanthe-
mum*, but gifted by Mr. Gosse with the appropriate
name of Beadlet, on account of the row of brilliant
azure beads which are seen just under the tentacles.

The colour of this species is exceedingly variable, and specimens may be found of almost every intermediate shade between olive-green and dark red. Brown is a very common tint; but even in the lower specimens a decided tinge of green or red is sure to be seen.

The Beadlets may be found by hundreds on any of our coasts where they can find a rock to which they can cling. They are among the hardiest of their race, and suffer no injury from a lengthened exposure to the air. Indeed, they may be taken in profusion upon rocks which are more than half-way between low and high-water mark, so that they spend a considerable portion of their lives out of water.

Plentiful as they are, they are not easily detected by an unpractised eye. When affixed to the rock and deprived of water, their multitudinous tentacles are withdrawn, the brilliant row of azure beads is invisible, and the creature that lately bloomed like a living flower resembles nothing more than a casual lump of

Beadlet, when closed.

green or red jelly that has been flung against the rock and stuck there. Its appearance when closed is shown in the illustration just given. A very short experience, however, will enable the observer to detect the Beadlet at a glance, and he will identify it as easily when it is closed as when it is blooming in all its wonderful beauty.

Not only is the colour variable in different specimens, but it has been known to alter in a single individual. More or less light, the density of the water, the supply of food, and other external conditions, have a wonderful effect in the colouring, and will force a Beadlet to gain or lose its tints within a comparatively short space of time. Still, however variable may be the colouring of the Beadlet, there is one point in which almost every specimen is alike. A narrow line of rich blue runs along the edge of the base, and adds much to the beauty of the creature.

The Beadlets are seldom found singly; and where one is seen, others may be expected. Generally, one large specimen is surrounded by a cluster of smaller animals, which are evidently its progeny, and which have never stirred from the place of their birth. The greater number are carried away by the tide, but a few generally succeed in planting themselves close to their parent, and in their turn produce a new colony. They are marvellously prolific beings, and even in a common aquarium a single specimen will produce a crowd of young, that stud every piece of rock and little stone to which they can cling. In their early stage they are very pretty little beings, though they are nearly transparent, and do not possess the beautiful colouring which adorns the matured animal.

Opinions are much divided as to the necessity of feeding Actiniæ when in an aquarium; some saying that the creatures only swallow by mechanical action, and do not digest the food which they take; while others state that unless they are fed, they first lose their beauty and then their lives. I incline to the latter theory, and have always fed my specimens, though sparingly, and not with beef or mutton, but with the kind of food which they are likely to obtain in the sea.

This species has a curious habit of detaching itself from the object to which it is clinging, inverting itself in the water, hollowing its base, and so forming itself into a simple kind of boat, just as do the fresh-water snails on a fine day. I have often watched the progress of this action, but our space is too limited to permit me to do more than recommend the young observer to examine it for himself.

As may be supposed from its mode of life, it is a singularly hardy inhabitant of an aquarium, and will permit very great liberties to be taken without seeming to resent them. Pure fresh water it does not endure, but it will flourish even when the water has been reduced by evaporation to barely half its ordinary volume; and there is a well-known anecdote of some specimens which were by mistake put into a jug of porter, and which survived after an immersion of fourteen days. Consequently, the young aquarium-keeper cannot do better than take this species as his trial essay, and study its habits before he attempts to keep the more delicate species, which are killed by the neglect of a few days, or even of a few hours.

There is a splendid species of Sea Anemone which is plentiful on some of our coasts, and which although common, and sometimes rather despised because it is so common, is far handsomer than many species which are much admired, not so much for their beauty as their scarcity. Its scientific name is *Tealia crassicornis*; but I always used to call it the " Crass." Mr. Gosse, however, in his valuable " British Sea Anemones," calls it the Dahlia Wartlet. This title is a very appropriate one, because the spread tentacles bear a really close resemblance to the flower of the dahlia, and the body is covered with little projecting warts, to which are attached fragments of stone and other objects.

If the Beadlet is only to be found with difficulty, the Wartlet cannot be detected without twice
the care needed for the other species.   It does not

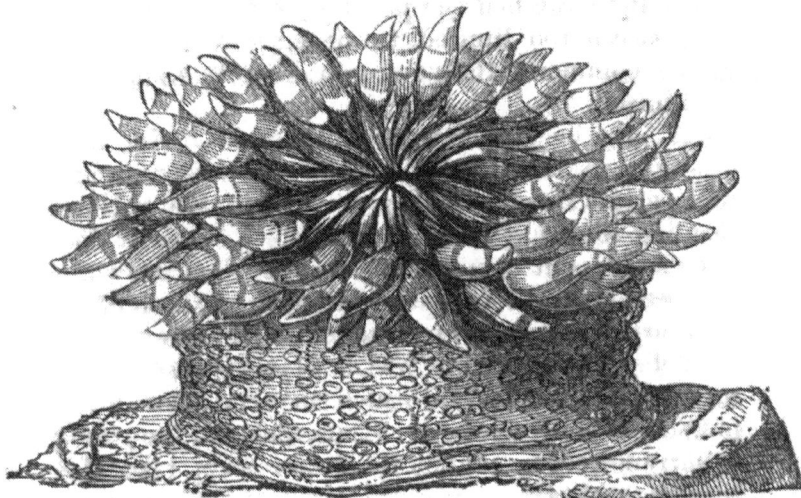

Dahlia Wartlet.

simply adhere to the rock, but prefers a place
where the rocks run well out into the sea, and are
partly covered with sand.   The wide base of the
creature is affixed to the rock, while it has the
power of contracting its body at will, so that it
is almost wholly buried beneath the sand.   And
as the body is thickly covered with pieces of shell
and stone adhering to the warts, it is hardly possible to detect a closed specimen until repeated
observations have taught the eye to mark the signs
of its presence.

Though so plentiful, it is curiously local, and
even in places so close to each other as Ramsgate
and Margate there is a marked difference in its
numbers.   On the shore of the former place it is
exceedingly plentiful, so much so that the bather
often recoils from the touch of its tentacles, and

three or four may be seen in the compass of a square yard. But off Margate it is by no means so common, and a very careful search must be made before it can be discovered. Moreover, as far as my own experience goes, it does not seem to attain the same size at Margate as at Ramsgate.

This splendid species takes almost any colour, from pearly white to scarlet and green. The tentacles are usually banded with crimson and white, and have a most lovely appearance when fully expanded. A fine specimen will measure more than five inches in diameter, so that an idea of its beauty may easily be formed.

Unfortunately, it is not to be recommended for the aquarium. In the first place, it is too large for the ordinary tanks; and, in the next, it is so very voracious. In its native state it eats green crabs, molluscs, shrimps, and, in fact, anything of an animal nature. It even captures the brisk and active prawns, arresting them in a moment, if they happen to touch any of its tentacles, and then forcing the unfortunate animal into its stomach.

Here I may mention that the tentacles are used in procuring food. In spite of their fragile and delicate appearance, they are most terrible weapons of destruction, armed with myriads of tiny arrows, and able to catch and retain even the strong and quarrelsome green crab. Sometimes this species obtains prey of a different character. Mr. J. Couch mentions that he saw a bee mistake an expanded Dahlia Wartlet for an open flower, and fly towards it. No sooner had it touched one of the simulated petals than it was arrested by the myriad harpoons that the tentacles can emit, and, in spite of its violent struggles, was gradually swallowed.

G

In the accompanying illustration are shown two figures of a pretty species of anemone that is much favoured by the owners of aquaria. It is called the Snake-locked Anemone (*Sagartia viduata*), and

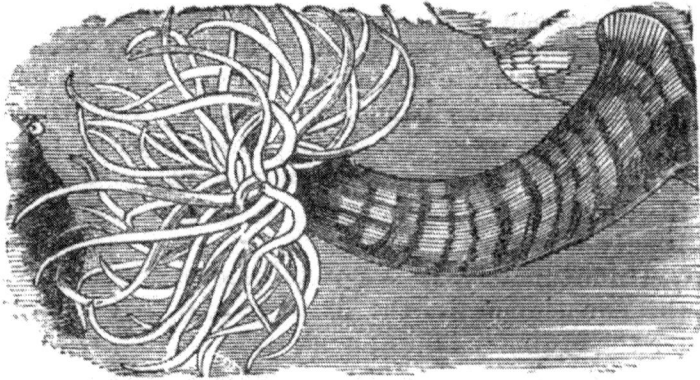

Snake-locked Anemone. Fig. a.

derives its name from the peculiar tentacles, which are very long, very slender, and very numerous, waving in the water like the fabled tresses of Medusa.

There are few species which exhibit more complete changes of form than the Snake-locked Anemone. Two of its changes are shown in the illustrations ; fig. *a* showing it as it appears when fully expanded, and fig. *b* as it appears when closed. I have always found it to be rather a capricious being. Sometimes it will become so flat that it is scarcely thicker than a sixpence, and every vestige of life seems to have vanished.

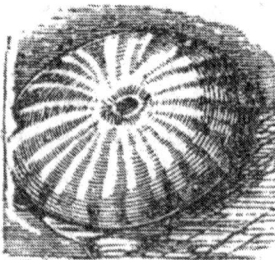

Fig. *b.*

Then if a supply of water be poured into the tank, or if it be roused from its inaction, it will rapidly swell out ; and in a minute or two, the flat plate of whitish jelly has become a splendid Sea Anemone, with a column several inches in length, and with some two hundred long tentacles waving gracefully in the water.

Though most plentiful on the Devonshire coast, this species is to be obtained in many parts of England. Mr. Gosse has mentioned a list of localities where it has been found, and I can add to that list the name of Ramsgate, where it was discovered by Mr. J. T. Hillier, who kindly presented me with some specimens. It is very local, and, as far as it is at present known, is only found in one place—namely, adhering to the woodwork of a sluice-gate. Still, as many young must be produced annually, it is probable that specimens may have taken up their habitation in many a spot at present unknown.

CHAPTER VIII.

SEA ANEMONES.

AS our space is limited, we must restrict ourselves to two more species. One of them is shown at Plate VI., fig. 2, and is called, from its peculiar aspect, the Plumose Anemone. Its scientific name is *Actinoloba dianthus.*

This species, when fully expanded, is undoubtedly the finest of its kind which inhabit the British coasts. The figure shows it as it appears when it is beginning to expand itself; but when it is fully extended, it is a singularly handsome creature, its column being sometimes six inches in length and three in width.

The colour of the Plumose Anemone is exceedingly variable. Sometimes it is pure white, sometimes slightly tinged with yellow, and, in fact, it is found of every imaginable tint between brown, yellow, white, and red.

The Plumose Anemone is chiefly remarkable for its puckered disc, which is divided into a number of frills that look almost as if they were delicate feathers attached to the animals. Sometimes the fringes are scarcely seen, and at others they are put forth in all their magnificence. Indeed, the Plumose Anemone is even more capricious than the Snake-locked species, and is continually modifying its form. As it dislikes light, and will not put forth all its glories except in darkness, the best mode of inducing it to expand itself completely is to cover the aquarium or to darken the room, and then the observer can come on it suddenly before it has had time to contract itself.

Our last species, the Cave-dwelling Anemone (*Sagartia troglodytes*), is chiefly remarkable for its free and roving habits. Sometimes it fixes itself by the base, protrudes its tentacles through the sand, and covers them with broken shells and small stones. A specimen is shown at Plate VI., fig. 3, as it appears when fixed, and in a partially expanded state. But sometimes it detaches itself from the object on which it has been resting, and allows the waves to carry it wherever they may choose. It will sometimes pass several months in a free state, the base being totally unattached, but the tentacles acting as anchors when it wishes to fix itself to any given spot. A figure of this species, when closed and unattached, is given at fig. 4 of the same plate.

We now pass on to another series of animals, called the Hydroid Zoophytes, several species of which are shown in Plate VI., figs. 5 to 12. All these species, together with others, were taken on the Ramsgate coast, and have been kept in an aquarium. These curious animals are very plant-like in their form, having stems, branches, and, in many cases, apparent flowers, which are, in reality, the animals by whom the stems and branches were formed. Indeed, for a very long time, these animals were thought to be seaweeds; and even at the present day, in those collections of seaweeds which are sold in marine watering-places, it is hardly possible to find one in which there are not several specimens of these zoophytes. Here I may mention that the term zoophyte literally signifies animal-plant; and, owing to the singularly plant-like form of many of the zoophytes, the name is very appropriate.

Our first example belongs to the genus Coryne, the species of which are known by the knobbed ends of the tentacles. A specimen is shown in Plate VI., fig. 5 ; and at fig. 6 a portion is drawn on an enlarged scale, for the purpose of showing the characteristic forms of the tentacles. Several species of Coryne may be found on our shores, and it is probable that the number of known species may be still more augmented.

At figs. 6 and 7 are shown two species of another genus, called Tubularia, on account of the tubular structure of the stem and branches, which are collectively known as the "polypary." Fig. 6 represents *Tubularia indivisa*, and fig. 7 *T. gracilis*. By the side of each is placed the polype on a magnified scale, in order to assist the observer in identifying the particular species.

Fig. 8 shows one of the commonest, as well as one of the prettiest, of British zoophytes. It is called *Sertularia argentea*, and looks, when in the water, like a beautiful feathery plume. Even when removed from the water, it is sufficiently firm and stiff to carry out the resemblance to a feather. A magnified portion is shown by its side. This zoophyte may often be found adhering to shells, and even to crustacea ; the common spider-crab being often so covered with it and these zoophytes that its shape is scarcely visible.

At fig. 9 is shown the elegant zoophyte called *Plumularia falcata;* and a magnified figure is given of a single branch, to show how the cells are set on the stem.

Fig. 10 represents a very curious zoophyte called *Laomedea geniculata;* and another species of the same genus, *L. gelatinosa*, is seen at fig. 11, each

being accompanied by a portion on a magnified
scale. The structure and—if such a term may be
used—the habits of the Laomedea are singularly
interesting; but even an imperfect description
would fill the whole of the pages devoted to marine
animals.

The last of our examples is one of the Bell
zoophytes, *Campanularia volubilis*, which is shown
at fig. 12. They have earned this name by the
bell-like form of their cells, from each of which
protrudes a tiny polype, with a number of delicate
tentacles. The whole of these zoophytes must be
examined by the aid of a magnifying-glass.

On Plate V., at figs. 4 and 5, are figures of two
common species of those creatures that are scientifically named Acalephæ, and are popularly known
by the name of Jelly-fishes. The latter term is
due to the gelatinous aspect of most of the species,
and the former is simply a Greek word, signifying
a nettle, and which was applied to the Jelly-fishes
in consequence of the stinging properties possessed
by many species.

Fig. 4 represents the beautiful creature which is
popularly called the Sea Acorn, and which has had
several names—that of *Cydippe pomiformis* being
generally used at the present day.

The animal is represented of the natural size, and
its colour is that of the purest glass, the tints of the
engraving being intended to imitate the iridescence
which plays over its surface at every movement.
In general shape the Cydippe is very like a melon,
and along the exterior run narrow bands, which are
seen to be ever changing their colour, all the hues
of the rainbow rippling over them in the most
wonderful manner.

A close inspection with the magnifying-glass
shows that these bands are composed of a vast

number of tiny flaps, very much like Venetian blinds on an infinitesimal scale. Each flap moves up and down in regular rotation, and the result is that the light is refracted in a variable manner, and the hues that are produced necessarily change with each movement of such flaps. There are eight of these bands in the Cydippe.

When the creature is floating in the sea, it is almost invisible, and even when placed in a glass vessel of water, a practised eye is needed to detect it. Indeed the only mode of doing so is by the faint and evanescent bands of coloured light which ripple over the animal as it moves to and fro in the water.

The best mode of catching this and similar inhabitants of the sea is by towing a fine net astern of a boat, letting the net come well to the surface. Every now and then, when the net is examined, it will be found to be studded with sundry little lumps of translucent jelly, apparently as destitute of form and life as of colour. These should be transferred to a vessel of water, by gently turning the net inside out, and placing it in the vessel, when the little creatures will detach themselves and swim away into the water, where they seem to disappear like magic. Sometimes, especially after a gale, myriads of them may be found upon the shore; but they are mostly dead; even those which have been flung into the high rock-pools seldom surviving.

The Cydippe is a very pretty inhabitant of the aquarium, and its movements are singularly graceful. It swims here and there with a ceaseless motion, turning over and over, trailing after it the long tentacles, which contract and elongate continually, while over its surface light bands of iridescent hues are continually rippling.

Neither this, nor any other Acaleph of which I

have personal knowledge, will live very long in an aquarium; but for the time that they can be kept they are most beautiful and interesting beings.

At fig. 5, Plate V., is shown a very pretty and tolerably common Acaleph belonging to the genus Sarsia. This is one of the innumerable creatures which we know by the name of Medusæ, in consequence of the long fibres which are trailed after them, and which writhe and twist about in the water like the serpent tresses of the mythical Medusa. Some of these Medusæ are gifted with terrible weapons, and by the aid of tiny fibres, not thicker than the thread of a spider's web, they can kill many of the smaller inhabitants of the sea.

Some of the tropical species are so poisonous that if their floating tresses coil round a human being they affect him as if he had been stung from head to foot by a swarm of wasps. In the colder seas of the British shores these formidable beings cannot live; but even on our own coasts a peculiarly venomous species exists. Its scientific name is *Cyanea capillata*, and it is popularly called the Stinger, or Stanger. After a southerly gale, this species is more common than agreeable, and I have even seen specimens floating up the Medway nearly as high as Upnor.

The poison-threads of this species are apparently interminable, and I have been stung by them when at a distance of many yards from the owner. Moreover, the threads sting as freely when separated from the Medusa as when attached to it; so that a single specimen may inflict an enormous amount of pain, if it should happen to float among a company of bathers.

The effect of the poison varies according to the individual. Some persons care little for it; and suffer scarcely any pain, except a sharp sort of pricking sensation, which is unpleasant at the time, but leaves no after-effects. Others, however, who, like myself, are afflicted with a sensitive skin, not only suffer severe and prolonged pain, but their health is seriously affected. In my own case the action of the heart and lungs is deranged for several months; sharp pains dart suddenly through the body, as if caused by a rifle-bullet; and occasionally these sudden pangs have been so fierce that I have dropped as if shot, and have been forced to lie on the ground for some time, before strength returned. I have known these symptoms recur more than six months after meeting with the Cyanea.

All the Medusæ which possess the bell-like disc, whatever its shape may be, propel themselves through the water by regular and almost rhythmical pulsations of the disc.

PART II.

—•◇•—

THE FRESH-WATER AQUARIUM.

WE now come to the Fresh-water Aquarium. The same directions as to the kind of vessel which is to be employed, the position, and general arrangement of the aquarium, apply equally to both. There are only two points in which great caution should be exercised.

In the first place, the sand, gravel, stones, &c., that are placed in the tank ought to be placed in boiling water for a few minutes, and then to be carefully washed and rinsed. If this precaution be not taken, various decaying substances, both animal and vegetable, are apt to adhere to the stones, and to cause unexpected havoc among the inhabitants of the aquarium. In the running streams, or even in the pond, these substances become innocuous, and are rapidly consumed by the myriad beings that people the waters; but in the narrow limits of an aquarium the order of nature is disturbed, and the balance must be maintained by careful superintendence.

In the next place, the water must on no account be drawn from a well or a pump, and, indeed, any kind of water that is used for drinking is unfit for the aquarium. The best water is that which is drawn from a river, and next to that is the water of a pond. Ordinary rain-water will, however, answer all purposes, provided that it be clear, and that the butt from which it has been taken has been provided with a cover. Hard water is injurious to all inhabitants of an aquarium, and to some it is an immediate poison.

The aquarium having been filled with water, the next point is to stock it. Although not absolutely necessary, a little vegetation has a very pretty appearance, and may be advantageously placed in the aquarium. Many fresh-water plants are suitable for this purpose, three of which are shown in Plate VII.

Fig. 4 represents the well-known Frog-bit, so plentiful in our streams. Fig. 5 is the Anacharis, that singular weed which has spread itself so rapidly through the country, and has made the pastime of boating almost impossible on many rivers. Fig. 6 is a plant which is not indigenous to this country, but may easily be obtained. Its name is Vallisneria, and the long, slender leaves have a very pleasing appearance in the aquarium. The manner in which this plant is reproduced, the tiny male flowers which float freely on the surface of the water, the female flowers at the end of their slender spiral stalks, and the singular movements of the plant, are most interesting, but do not come within the scope of these papers.

We will now proceed to the living inhabitants of the fresh-water aquarium.

The two figures, 1 and 2, on Plate VII., represent the male and female of the common Newt (*Triton cristatus*). This animal, which seems to belong to the lizards, is in fact one of the same group as the frogs and toads, and, as the reader may perhaps remember, looks something like the tadpole of the frog after it has put forth its legs, and before it has lost its tail.

The Newt, Eft, or Evat, as it is indifferently named, is very plentiful in this country; and there

is scarcely a pond or an old-established ditch that
is not tenanted by it. Should the reader like to
take some specimens for the aquarium, he can do
so by two methods ; namely, angling for them, or
catching them in a net. They are very fond of
worms, and if a thread be tied round the middle
of a worm, and the bait be thrown into the water,
the Newt is tolerably sure to grasp it, and may be
coaxed to the edge, and then drawn out on the
bank before it loses its hold. A common ring-net
will, however, answer perfectly well, and if judiciously used, any number of Newts can be taken.

Many persons have a strong antipathy towards
the Newt, an antipathy which is entirely without
foundation, inasmuch as it is one of the most
harmless animals that can be conceived. Yet in
most parts of England the Newt is hated and
feared as much as if it were a rattlesnake, and
there are scarcely any bounds to the extravagant
stories that are told about its noxious powers. It
is poisonous, it is ferocious, it has venomous teeth,
it bites cattle, and it spits fire. This last accu-
sation—rather an absurd one, considering that the
animal lives in the water—is perhaps more widely
made than any other.

Yet the Newt is perfectly harmless, and it is
very pretty. The male, during the breeding season,
may lay claim to special beauty. The colour of the
upper part of the body is olive-green, mottled with
dark brown ; the under parts are rich orange, with
black spots, and along the back runs a most delicate
membranous fringe, deeply serrated, like a cock's
comb, and being of a lovely scarlet hue. This
appendage is only worn for a short time, when it
is absorbed into the body and disappears, leaving
scarcely a trace behind it.

If possible, the aquarium-keeper should watch

the whole life of the Newt, from its first appearance in the world, in the shape of an egg tied up in a slender leaf, through all its changes of structure, until it has attained its perfect form. The Newt will change its skin several times, and the cast skin should be preserved for examination. It is very thin, and floats in the water like a film. It can be removed by slipping a piece of white cardboard under it, and arranging the skin upon it exactly as seaweeds are fixed on paper. No cement will be needed, as it will fix itself to the cardboard so firmly that, when dry, it seems rather to be a drawing in pale sepia than a veritable integument.

We will pass over the frogs and toads, and, in our next, come to the fresh-water fishes.

CHAPTER IX.

FISHES.

AT fig. 6, Plate VIII., is shown a figure of the Three-spined Stickleback, one of the best-known inhabitants of our inland streams. Ditches are especially favoured by the Stickleback, particularly if a tolerably swift current runs through them. The Sticklebacks are notable for the peculiarity from which they derive their name, *i. e.*, a number of sharp spikes projecting from the body. The number of these spines varies in different species, some having as many as fifteen, and others, as in the present case, only three. Their bodies are not wholly defended with overlapping scales, as in the generality of fish, but are protected by rows of bony plates, between which the bare skin of the creature is visible.

In many species the skin of the males glows with the most brilliant colours,—red, green, and gold being the most conspicuous ; and in the breeding season the appearance of these little fishes is truly beautiful. Indeed, if they could only be magnified some twelve or fourteen diameters, they would not be excelled in beauty by any inhabitant of the sea or river, and not even the gorgeous fishes of the tropics would surpass them in brilliancy and richness of colouring.

When the Sticklebacks are kept in the aquarium, they must either be placed in a large tank or in separate vessels, as they are quarrelsome to a degree, and can fight to the death with their spiked weapons.

They will fight on any pretext, and without any pretext at all, and in the fighting season a couple of males cannot meet without a skirmish. But the objects for which they generally fight are two—their wives and their homes. That they should fight for the possession of the former is not extraordinary, as there are few animated creatures in a wild state who win their mates without a fight, and fishes are not exempt from this general law.

But there are few fishes that possess veritable homes like those of the Stickleback. They may choose certain portions of the beach for their residence, or may take up their position under some convenient rock, but the Stickleback is one of the few fishes that not only possess a house, but build it.

The nest of this curious little fish is made of vegetable fibres, which are woven together with sufficient strength to form a definite nest, in which the eggs can be deposited. In shape the nest is something like a small barrel open at both ends, so that the fish can enter at one end and go out by the other. It is a small nest, and does not conceal the fish that inhabits it ; for as the Stickleback lies in its little house, the head projects at one end and the tail at the other.

The reason why the Stickleback is obliged to make a nest is simple enough. It is well known that there is no food of which fish are so fond as the eggs of fish, and that in the case of fishes that produce myriads of eggs a very small percentage is hatched, and a still smaller reaches an adult age. But the Stickleback produces but a few eggs, and those of very great comparative size. One Stickleback egg is equal to some twenty or thirty eggs of the Codfish, and is particularly liable to be eaten. by the other inhabitants of the streams, and by none more voraciously than by the Sticklebacks themselves. If a nest be opened, and the eggs thrown singly into the water, the result is most amusing,—all the Sticklebacks in the neighbourhood darting to the spot, converging upon the egg, and fighting furiously for it.

It is therefore necessary that the eggs should be guarded in some manner, and this is done by placing them in the nest, from which the male fish never stirs except to procure food and to fight, and in no case does he venture far from his home. The females do not trouble themselves about their future young, and, indeed, as a number of females generally deposit their eggs successively in the same nest, it would not be likely that they should do so. The male, however, is quite equal to the task, and is so very fierce during the time of his guardianship, that he will allow nothing to pass within the range of his domains without attacking it.

When another male happens to approach, there is always a combat, each fish trying to force its way under the other, so as to inflict a wound with its spikes on the unprotected portions of the enemy's belly. During the fight, the green, gold, and scarlet of the little fishes glow with double richness; and when one of them is beaten and driven off, the conqueror seems to increase in beauty, while the colours of the vanquished fade perceptibly, and in some cases almost wholly vanish.

The nests of the Stickleback may easily be taken from the water by means of a ring net, and placed, together with their guardians, in the aquarium.

On Plate VIII., fig. 8, is a very odd-looking fish with a large flattened head, goggle eyes, and a wide

H

mouth. This is the little fish that is popularly known by the various names of Bull-head, Miller's Thumb, and Tommy Logge. Its scientific title is *Cottus gobio*.

I have selected this little fish because it exemplifies a well-known principle in nature,—that form is always adapted to situation. On looking at the fish, we see that its head is very wide and very flat, the substance of the head running out at the sides in proportion to the flattening above. Now, what does this mean? Why is the head flat, and, being flat, what does the flatness portend?

The practical naturalist can answer all these questions with ease and certainty, even if he had never seen the fish before; if the Bull-head were a recently discovered species, he could tell much of its habits by only looking at the head.

He knows that form is always adapted to conditions. He knows that the Mole, the Eel, the Worm, and the Ferret, are all furnished with long, slender bodies, because they have to insinuate themselves through small apertures. He knows that the swift of foot may be known by the development of their legs, and the swift of flight by that of their wings. The paw or tooth of the tiger tells him the rapacious character of its owner; while the wide pillar-like foot of the elephant indicates the enormous weight which has to be supported on such a basis.

Now, it is an invariable rule that wherever a crevice is found, there is also found some flat creature that can creep into it. This is exactly the case with the Bull-head, which loves to hide under stones—a habit for which its wide, flat head renders it particularly fit. Consequently, it is chiefly to be found in those parts of rivers where the bed is stony, and in all such localities it is very sure to be plentiful.

It is a curiously apathetic fish, though it alternates intervals of long repose with a few moments of swift and energetic action. Generally, it lies at the bottom of the river, with its body, or at all events its head, concealed under a stone, and will remain there so quietly that an unaccustomed eye would fail to perceive it, even if the precise spot were pointed out. But if the stone be removed, the Bull-head waits for a second or so, as if in wonder at the change, and then darts off with a speed and suddenness that seem scarcely compatible with its ordinary sluggish habit. Probably it is acquainted with every hiding-place in the neighbourhood, as it darts from one place of refuge into another with such rapidity that it seems to vanish as if by magic, and it is very difficult to tell under which particular stone the little fish may have taken refuge.

Closely as it lies, however, it can mostly be discovered by a practised eye, which, indeed, can see objects which are quite invisible to ordinary visions. There is scarcely any animal, however cunning, or however well furnished by nature with the means of concealment, that can hide itself from the experienced hunter; for the creature has life, and life will always betray itself. The animal must at least breathe, and even the slight movement caused by respiration will be seen by those who know what to look for. Now, the Bull-head betrays itself in a very peculiar manner. It has an inveterate habit of wagging its tail, and even if the searcher after the fish cannot see the tail itself, he can at least see the current of water that is caused by the action.

Just as the flattened head of this fish shows that it is in the habit of lying under stones, so does its long and wide mouth indicate that it is a voracious species, and capable of eating creatures of considerable comparative size. Its usual food consists of worms and various aquatic insects, whether in the larval, pupal, or perfect stages of existence.

It is also a great enemy of the small and helpless fry of various fish, and can on occasion devour a fish of considerable size. It has been known to catch and devour a Minnow of full size; and as the adult Minnow is really as long as the adult Bullhead, the voracity of the fish may be easily imagined. As all the voracious fish have a very quick digestion, it is possible that the partially swallowed fish may be digested gradually, so as to allow it to descend by degrees into the insatiable stomach of the devourer.

At fig. 2 of the same plate is the well-known Perch (*Perca fluviatilis*), which is the acknowledged type of an enormous family of fishes inhabiting both the salt and fresh water.

The owner of an aquarium can scarcely have a handsomer fish for his small aquatic world; and every one who sees a living Perch must admire the rich green back crossed by its dark bands, the golden white belly, and the 'scarlet fins.

This is a bold and voracious fish, and will eat almost anything that may be given to it; preferring, however, worms, insects, and small fishes.

One of the most interesting points in the Perch is its wonderful egg-ribbon. This curious object is sometimes from four to five feet in length, and if carefully examined in water, is seen to be not a mere ribbon, but a complete tube. The eggs are arranged with beautiful regularity; and, on account of their pressure against each other, become rather flattened at the points of contact, so as to bear some resemblance to a honeycomb.

As the egg-ribbon is many times the size of the fish which laid it, the young observer naturally feels puzzled at so curious a fact, and needs explanation of an apparent impossibility. As long as the eggs are contained within the body of the parent fish, they are of very small size, and are closely packed together. Any one who has seen the "hard roe" of a herring or any other fish, will at once form a good idea of the size and shape of the eggs, and the manner in which they are packed.

As long as they are contained in the fish, they are kept from the water, and retain their shape and size, but as soon as they are deposited, and the water has free access to them, a very remarkable change takes place. Each egg is surrounded with a sort of gelatinous envelope, and as soon as the egg-ribbon is laid, the gelatinous envelope attracts the water, which passes into it by a curious process which is technically named "endosmosis," and in consequence each egg swells to many times its original extent. In looking at the egg rather carefully, the observer will see that the germ of the egg, which was all that was perceptible in the "roe" state, has scarcely altered at all in shape or size, but lies in the middle of the spherical gelatinous mass with which it is surrounded.

A similar phenomenon takes place with the spawn of the frog and toad, and, especially in the case of the former animal, produces most singular results. Few persons can be made to believe that the huge mass of spawn which floats in the ponds and ditches can be produced by single frogs; and the general idea is that each such mass is formed by a number of frogs which have chosen to lay their eggs in the same spot.

Keep your Perch alive as long as you can, and

when it dies, make the best use of its body, and examine it in detail. You can hardly have a better example of the structure of fishes than is found in the Perch, which has been chosen by several of the most eminent anatomists as affording an excellent type of the curious structures which they wish to investigate.

Examine the wonderful structure of the gills. Remove them from the head, place them under water, and examine them with a magnifying-glass. You will be surprised at the enormous amount of surface which they present to the water; and if you have a glass of sufficient power, you will be able to see on the surface the minute veins and arteries which bring the blood in contact with the oxygen contained in the water, and so enable the fish to live.

There are few objects that cause greater surprise to a novice than the gills of a fish when opened out, magnified, and explained.

Perhaps I may spend a few lines on the use of this beautiful structure. Every one knows that the gills are used for breathing, but it is not every one who knows how they are employed.

The blood of all breathing animals becomes deteriorated in its passage through the body, and requires to be re-vivified by contact with oxygen before it is again capable of doing its work. In highly organized beings, such as man, the renewal must be carried on with great frequency, or death soon follows. If, for example, the respiration of a human being is interrupted but for three minutes, death is sure to come, unless potent and scientific remedies are used. The heart may still beat, and the blood still circulate, but its life-giving power is gone. In the unvivified condition it absolutely becomes a poison instead of a vivifier;—

the brain is paralyzed, giddiness ensues, followed by insensibility, and then by death.

In the higher beings, the oxygen which is contained in the atmosphere is brought in contact with the air in a very beautiful manner, and in the act of breathing a very difficult problem is solved.

How is it possible to allow the atmosphere to come in contact with the blood, without permitting it to escape from the vessels through which it flows?

This difficulty is surmounted in a very simple manner. A portion of the circulating apparatus is so contrived that the vessels are diminished to the minutest possible size; so small, indeed, that there is only just room for the globules of the blood to pass singly. The walls of the vessels are extremely thin and delicate, and are made of a substance which allows the passage of air while it retains the blood. In the larger vessels, such a structure would be impossible, on account of the pressure to which the walls are subjected by the volume of blood that rushes through them; but in those parts which are exposed to the air, the currents are so minute that they exercise comparatively little force, and are easily contained within their delicate walls.

It is, of course, an important point that a very large surface should be exposed to the air, and it is hardly possible to find a better example than the gills of a fish. At first sight they look like a series of comb-like organs, scarlet with the blood that is seen through their delicate coverings; but if they are closely examined, they will be found to possess a most beautiful form, exposing a very large surface, and at the same time occupying a very little space. Each tooth of the scarlet comb is composed of innumerable plates of membrane, traversed by the blood-vessels, and admitting the air on both sides. In fact the gills remind the observer of the leaves of a slightly-bound book, in which a very large amount of surface is compressed into a very small space.

I have been particular in describing these gills because they teach the young observer the real action of respiration better than any structure that can be found. Moreover, they can easily be obtained, and an ordinary magnifying-glass is sufficient to exhibit their wonderful mechanism.

The manner in which a Perch breathes is simple enough. The fish opens its mouth and admits a certain quantity of water, just as we admit air into our lungs. It then closes its mouth, and drives the water out at the gill-covers, causing it to wash over the gills in its passage. The oxygen contained in the water thus comes in contact with the blood, and so the fish manages to breathe.

CHAPTER X.

FISHES (*continued*).

THE reader will now understand why a fish always lies with its head up the stream. By so doing it enables the water to pass through its gills without any exertion being required. Also, the reader will understand why an angler who has a strong and heavy fish on his line tries to lead it gently down the stream. By so doing he greatly hinders, even if he cannot entirely prevent, respiration, and thus weakens the creature, which cannot exert itself without a sufficient supply of breath any more than a human being can. In fact a skilful angler can nearly drown a fish in its own element.

Having looked at the gills, lay the head quite open, and look for the brain. You will be quite surprised when you see it. It is singularly small in proportion to the size of the fish. Not being able to procure a Perch in this part of the country, I have just opened the head of a Whiting, and find that the brain is not larger in proportion to the volume of the body than a walnut would be in proportion to an ordinary-sized man. In the specimen just mentioned the brain is scarcely one-third as large as one of the eyes.

Now lay the fish open from the chin to the middle of the body, and you will see what a very small space is occupied by the vital organs. No room is wanted for the gills, which are disposed of elsewhere, and the rest of the organs are small and simple. Look at the little heart, which lies just under the throat, and which is only composed of two cavities, whereas that of the higher animals has four. Next, examine the swimming-bladder, that curious piece of mechanism which lies just under the spine, and which enables the fish to rise or sink at pleasure, by compressing the air within the swimming-bladder, or allowing it to expand.

Now look at the scales with which the body is covered. Perhaps the reader may wonder how the scales grow in exact proportion to the size of the fish. If he will examine one of them with the magnifying-glass, he will see that they increase by adding new matter at the edges; so that each scale increases exactly in proportion to the growth of its owner. The numerous concentric lines on the scales mark the growths of successive seasons, just like the rings in timber. There is one point in the scales to which particular attention should be directed. If you lay the fish on its side, you will see that there is a conspicuous narrow line which runs from the gill-covers to the tail, and which nearly follows the arch of the back. This is called the "lateral line," and its shape is of great use in distinguishing one fish from another. It is formed in rather a curious manner. Each scale of this line is pierced near its base with a little hole which corresponds to an aperture in the body of the fish. Through this aperture is poured that slimy substance with which the scales of the fish are covered, and which serves as a defence against the water.

I am urgent for the young observer thus to take a fish to pieces, because he will in this manner learn more of fishes in an hour or two than he will do by many days of mere book-work.

In England we value the Perch for two purposes only,—namely, for food and for sport. In Lapland, however, it has another use, the skin being employed in the manufacture of bows. These weapons are made of several pieces of wood united together by the skin of the Perch. A sufficient number of skins are deprived of their scales, and are then bound tightly round the bow with birch bark. The whole is then boiled for a considerable time; and when the weapon is suffered to dry, the wood is firmly united by the perch-skins.

The young aquarium owner may perhaps be glad to learn that this fish can be tamed without any difficulty, and that specimens have been rendered so docile as to come and take food out of the fingers.

Next in order comes a fish of which the aquarium-keeper must needs be very careful. This is the Pike, or Jack (*Esox Lucius*), a fish which has aptly been called the Shark of the fresh water. And so it is, as far as England is concerned; though there are fresh-water fishes inhabiting other countries that are far more shark-like in their habits, and even become dangerous to men and cattle. (See Plate VIII., fig. 4.)

The voracity of the Pike is well known, and I need hardly mention that if any other fish is placed in the same receptacle, the Pike will assuredly eat it, if it is not very much too large. Even if it be so large that it cannot be all accommodated at once, the Pike will be happy to swallow as much of it as the stomach will hold, and trust to the process of digestion for enabling the remainder to be swallowed by degrees. As to little fishes, the Pike will dart at them in a moment, seizing them by the middle, carrying them to its abode under a stone, or some similar locality, and then swallowing them whole.

It is said that there are two fish which the Pike will not touch ; namely, the Perch and the Golden Carp, properly called the Gold-fish. The Pike is reputed to be afraid of them, the strong prickly spines of the Perch deterring him from attacking so well-armed a prey. And, in proof of this assertion, it is said that if an angler has been unsuccessfully attempting to catch a Pike, he can mostly succeed by taking a rather small Perch, cutting off the prickly fin and using it as bait. The Pike seeing that the Perch is defenceless, will not lose so excellent an opportunity : he accordingly darts at the bait, and is straightway hooked.

The Gold-fish is said to frighten the Pike by reason of its bright scales, with their metallic radiance. The voracious fish has never seen anything of the kind in English waters, and dares not attempt the capture. However, I should be very sorry to trust either a favourite Perch or Gold-fish in the same aquarium with a hungry Pike, especially if the latter were much superior in size. Indeed, the Pike has been known to swallow a Gold-fish under such circumstances, much to the astonishment of the owner, who had previously held the opinion that the two creatures could live as safely together as a couple of lambs.

The Pike will eat its own kind quite as freely as any other fish, and if possible, seems even to prefer a little cannibalism. Many experienced anglers have narrated instances where Pike of some pounds weight have been found in the interior of larger fish of their own species.

The Carp (Plate VIII., fig. 3) is not very often kept in aquaria, its place being easily supplied by the Gold and Silver Carp brought from China. However, even our own fish is well worth keeping, as in a good specimen the large broad scales are tinged

with a golden hue that shows out well in the water.
It is a hardy fish, and as it is able to live for a
considerable time out of water, provided that its
gills be kept moist, it can be removed from place
to place if wrapt carefully in wet moss. Indeed,
in some parts of Europe, Carp are fattened for the
table by being kept in nets filled with wet leaves,
and being constantly fed with a paste made of bread
and milk.

One of the handsomest of British fresh-water ·
fishes is the common Trout (Plate VIII., fig. 1),
whose crimson-spotted sides always excite great
admiration. It is not nearly so hardy a fish as the
Carp, and requires considerable care to preserve it
in good health. It can be fed with worms, little
minnows, or even with large fish, if it be of any
size.

The Eel is always welcome to the aquarium-
keeper, as being well adapted for living in so small
a house. I have kept many of them, and have
often wished that other fish gave as little trouble.
(See Plate VIII., fig. 7.)

One, in particular, about six inches in length, I
kept for some seven or eight months in a common
Gold-fish globe. It was perfectly healthy, and
occasionally extremely lively, undulating its way
through the water with an easy grace that belongs
to no other creature. Although in perfect health,
it did not appear to grow ; at all events, the increase in size was so trifling that it was hardly
perceptible. Why such should have been the case,
I do not know, as it took food, and appeared well
contented with its situation. At the end of that
time, I transferred the fish to the fountain of one
of our largest hospitals, and never saw it more.

A large Eel is, of course, useless in an aquarium,
chiefly because it takes up so much space ; but a

small one is a pretty creature, and is besides useful in exhibiting a curious anatomical structure. Take a little Eel, and wrap it up in a wet cloth, only allowing a little piece of the tail to appear. Now, lay the tail on a flat piece of glass, cover it with water, and examine it by means of a microscope. In default of an elaborate instrument, a common magnifying-glass will answer the purpose. A pulsating vessel will then be seen at the end of the tail, beating as regularly as the heart, and evidently acting in that capacity towards the liquid in which it acts.

While we are on the subject of the Eel, I will make a few remarks on the origin and spread of the fish.

Until within the last few years, the wildest and most contradictory ideas were prevalent regarding the origin of the Eel. That the Eel deposited eggs like other fishes was discredited, because no one had been able to find any roe at any period of the year. A curious variety of theories were then set afloat. The most popular was, that the Eels were originally hairs from the tails of horses. These fell into the water when the horses came to drink, and by dint of long soaking became endowed with life, and turned into worms of the same shape as the original hair. As these worms grew, they assumed a head at one end and a tail at the other, fins grew in the proper places, and by degrees they became genuine Eels.

This theory sounds absurd enough, but it is really not worse than many others. In almost all of them the "Hair-Eel" is accepted as the origin of the real fish, though the origin of the Hair-Eel seems to be as ill-understood a problem as that which it is intended to replace. The Hair-Eel, or Hair-worm, in reality belongs to the great tribe of worms, and is scientifically known by the name of *Gordius aquaticus.* Its first, or generic name, is given to it because it twists its long and slender body into folds as complicated as those of the celebrated Gordian knot, and its second, or specific name, is earned by its habit of living in water. It may often be found in pools and sometimes even in puddles, gathered together in a small compass, and being so long, so hard, and so slender that we really cannot wonder that the ignorant rustics believe it to be an animated horse-hair. There is another locality where the Gordius may be also found, though few would look for it in such a place. Every one is familiar with the black, shining, active beetles that are so plentiful in the country during the summer months. Several species of these beetles become the home of the Gordius, which lives in the interior and packs itself up quite wonderfully in so small a space. I well recollect my surprise at finding out this fact. Shortly after my introduction into entomology, I took some of these beetles, and put them into spirits of wine for the purpose of preserving them. My astonishment was very great, when I saw a Gordius issue from one of the beetles, and protrude itself to a considerable length. The spirit was so fatal in its effects that the Gordius could not extricate itself entirely, but died when partly in and partly out of the beetle.

The creature frequently issues from the beetle in water, whither a kind of instinct seems to drive the insect; and this is the reason why a puddle which was only caused by recent rain, will sometimes harbour one or two of these curious worms. On this fact, so well known to naturalists, a theory has been founded, and even promulgated in type, to the effect that the Eel was originally the

intestine of the beetle, which became animated
when the insect died, then turned into a hair-worm,
and thence was developed into an Eel. And as a
proof of the truth of the theory, this familiar fact
was cited that Eels are found in ponds where no
such fish existed a week or two ago.

Now, every fisherman knows that Eels, like
many other fish, have the power of sustaining life
for a long time after they have been taken out of
the water, and also, that their peculiar serpentine
movements enable them to travel over land for
considerable distances, especially if the ground be
wet; I have even found them crawling up a
perpendicular rock, and caught several specimens
while they were actually working their way along
a stone that overhung the water. They seemed to be
held to the stone by the pressure of the atmosphere,
just as a piece of wet paper will stick to the wall.

All the previously-mentioned theories, together
with many others, are now exploded by the power
of the microscope, which has discovered that the
Eel has spawn like that of other fishes, only that
the eggs are of very minute dimensions. The so-
called "fat" of the Eel is really the spawn, or roe,
the individual eggs being too small to be distinguished by the naked eye, though they are easily
detected by the microscope.

Our last example of the fresh-water fishes is
the Lampern, or River Lamprey (*Petromyzon
fluviatilis*).

This curious little creature is closely related to
the well-known Lamprey of the sea, but it is not
known to leave the river for the ocean. In some
rivers of England it is exceedingly plentiful, and
may be seen wriggling its way up the stream, or
clustering in thick masses upon the stones of the
river-bed, sometimes being so numerous as to look

like great bunches of loose weeds. To these stones
the Lampern adheres by means of the mouth, which
is made precisely on the same principle as that of
the leech, the circular lip forming a sucker from
which the air is exhausted. Within the lips are

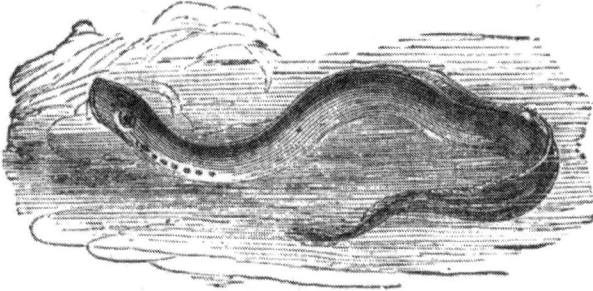

The Lampern.

teeth, whose points curve towards the centre of the
mouth ; so that by moving the jaw within the lips,
any soft substance can be bitten into and scraped
away.

The ordinary food of the Lampern is said to
consist of aquatic insects and worms, and the flesh
of dead fish. But within the last few years several
letters have been sent to the various journals which
treat on Natural History, stating that the creature
had been observed in the act of attacking living
fish, and gnawing circular holes in their bodies.
Even the Salmon has been hooked with the
Lampern attached to its body, the fish having
already scooped a shallow hole corresponding with
the size of its mouth.

This sucker-like mouth is also used for another
purpose,—namely, the formation of convenient
beds in which the eggs can be deposited. It is
very interesting to watch the fish engaged in this
process. Having fixed upon a suitable spot, it

I

will affix itself to a stone, and then swim back
wards down the stream, pulling the stone after it,
and making the most extraordinary evolutions in
effecting its purpose. After dragging the stone for
a yard or two, the Lampern drops it, and then
returns for another stone. In this manner it
proceeds until it has scooped a shallow hole some
eighteen inches in diameter.

Now let us examine some of the leading characteristics of this curious fish.

The Lampern has no gill-covers, but the water
obtains access and egress in a very peculiar manner.
While it is adhering to a stone, the mouth is closed,
so that no water can pass to the gills. But it is
quite independent of its mouth for respiration, for
on the top of the head there is a small round hole
which leads to the breathing organs, and which
admits the water to the gills. It obtains exit by
means of seven circular holes at each side. In
consequence of this structure, the fish is in some
places called the Seven-eyes, and in others the
Nine-eyes, the latter name being derived from the
seven breathing apertures on each side, the real eye,
and the aperture on the head, which is reckoned
twice over, once for each side.

The flesh of the Lampern is excellent, and is
as good as that of the Eel, though in many places
there is a violent prejudice against it. Its chief
use in this country is for bait, the Turbot and Cod
being easily attracted by it.

Before concluding this account of the fishes, I
should like to mention a fresh-water aquarium of a
singular character.

One of my friends, who lives in one of the
densest parts of London, takes his guests into a
little back room, where, to all appearance, the
inmates are partly under water, as if in a diving-

bell. There is only one window to the room, and that window is apparently the only means of keeping the water out. Through the panes are seen fish swimming about at their ease, sometimes sailing steadily along, and sometimes putting their noses against the window, as if trying to enter the room; aquatic plants are waving their flexible leafage in the water, while many other inhabitants of the river are flitting about as confusedly as if in their native haunts. In the middle is a fountain, which throws jets of water high into the air, while, as the spectator directs his gaze upwards, he seems to be looking into a nymph's cavern, rich with shells, stalactites, and glittering crystals, and lighted from above by the blue sky.

How this curious and beautiful effect can be produced, is not easily seen until the inventor throws up the window. As he lays his hand on the sash, the spectator is rather startled, because, to all appearance, the glass panes form the barriers against the water. However, the sash glides up easily, and the water does not come in. A closer view betrays the deception, which is really an ingenious as well as a pretty one. The aquarium is built just outside the window, and is about eighteen inches wider on either side. Both sides and the back are made of brick and slate, well cemented, while the front is of a single sheet of plate glass which is close behind the window panes, and is not seen when the sash is down. The tank is, of course, a very large one, and the back being about six feet high, and skilfully modelled into the semblance of a rocky cavern flooded with water, the whole arrangement gives the room a most unique appearance, because the inmates seem to be inhabitants of the cavern, and to be looking through the water at the sky.

CHAPTER XI.

FRESH-WATER MOLLUSCS, CRUSTACEA, AND SPIDERS.

NEXT in order we come to the Fresh-water Molluscs, of which we shall give but a few familiar examples.

In the first place the bivalve molluscs are of very little use in an aquarium, and are so apt to die and damage the water that they are not recommended. Of course, if the observer wishes to watch their habits, he can place some bivalves in separate vessels, but it is always a hazardous experiment to allow them to be in the same aquarium with other creatures.

We will take in regular succession the shells which are given in Plate IX.

Fig. 5 represents the pretty shell called *Bithynia tentaculata.* This shell is plentiful in our ditches, canals, and slow-running streams.

The reason why it is here introduced is, the curious and interesting manner in which it deposits its eggs.

It begins by looking out some smooth surface, such as the stem of a plant, a stone, or even a submerged leaf. The next process is to clean the surface very carefully with its mouth, and then to deposit an egg upon the cleaned spot. A fresh space is next prepared, another egg deposited, and so the creature continues to act until it has laid all its eggs.

Sometimes there are sixty or seventy of these eggs, which are globular in shape, and look as if they were little shot made of some gelatinous substance. They are arranged very carefully in regular rows, and the process of laying and fixing them occupies several days. As those molluscs are apt to affix their egg-bands to the glass of the aquarium, the magnifying glass can be used to great advantage, and even a microscope of moderate power—say half-inch focus—can be applied to them successfully, so as to observe the gradual development of the young.

The eggs are laid throughout the whole summer, from May to August, so that the observer will have many opportunities of watching the eggs and the mode in which they are hatched. The shell is represented of the natural size. Its colour is like that of horn.

At fig. 2, Plate IX., is seen a figure of a very common fresh-water mollusc, which the reader will probably recognise as soon as he sees it. This is the familiar water-snail, *Paludina vivipara.* This species is also shown of the natural size.

Closely resembling the preceding mollusc in many of its characteristics, it has one essential point of difference ; namely, that the young are hatched before they are sent into their watery world. There are about twenty or thirty of the young, and they are retained within parental protection until they have been hatched for two months at least, and have become able to find their own nutriment.

When they issue from the protecting shell, they do not make a simultaneous move, but issue forth one at a time, and so deliberately, that in the course of a whole day only three or four will have escaped.

This shell is very plentiful in the southern parts
of England. The Cherwell is thickly populated
with Paludinæ, and within a few miles of its
junction with the Isis at Oxford, there are favourite
spots where almost any number of them may be
procured. They specially seem to affect little bays,
where the water is comparatively still, especially if
it should happen to be protected by reeds, flags,
and other aquatic plants. The still water just above
"lashers" is almost always a favourite spot with
Paludinæ, and in the little quiet nooks and covers of
such places, the molluscs are sure to be found.

The colour of the shell is brownish-green, bounded
spirally with deep reddish-brown.

We now come to the flat shell seen at fig. 1,
Plate IX. This shell instantly reminds the
observer of that familiar firework called "Catharine
Wheel," and indeed, if this shell were charged with
the proper mixture, and a pin run through its
centre, it would probably revolve and throw out
its showers of sparks as well as if it had been made
especially for the purpose. Of course, it would be
but small, and its effect would correspond to its
size, but it would certainly produce a revolving
shower of sparks.

This shell is known by the name of *Planorbis
vortex*, and belongs to a genus which is rather
widely represented in England, a goodly number of
British species being acknowledged. Some of them
are exceedingly small, the largest examples not
exceeding the eighth of an inch in diameter, while
others measure as much as an inch in diameter.
The species which has been selected as our example
is very flat indeed, and is about one-third of an
inch in diameter. The substance of the shell is
thin and horny, and its colour corresponds with its
texture.

At fig. 6, Plate IX., is seen the shell of another common fresh-water snail, called *Limnæus stagnalis.*

This elegantly-formed shell is to be found in similar localities to those which have been already mentioned, and is quite as plentiful.

The genus to which this mollusc belongs is a tolerably large one, and six or seven species are known to inhabit England.

The eggs of these molluscs are placed close together, and are attached to stones or aquatic plants, but they are not like those of the Bithynia. They are more in number, and they are collected in elliptical masses of a gelatinous substance, which is protected by a delicate membranous envelope. Although these egg-masses cannot be so favourably viewed with the microscope as those of the Bithynia, they can be easily removed from the object to which they adhere, so as to be brought within the range of the magnifying-glass. Like others belonging to the same genus, this species is rather variable both in colour and in form. Generally, however, its shell is horn-coloured, and the length is about an inch and a half.

Our last example of these molluscs is the curious species which is called the Fresh-water Limpet, *Ancylus fluviatilis.* That it does not really belong to the Limpets need scarcely be mentioned, for although in the general shape of its shell it bears a strong resemblance to some of the marine Limpets, it is closely allied to the molluscs which have just been mentioned.

It is a pretty little creature, and is always an ornamental addition to the fresh-water aquarium. The general shape of the shell can be seen by reference to fig. 3, Plate IX.; but specimens as large as the figure are not often found, its length scarcely exceeding a quarter of an inch.

Owing to the small dimensions and inconspicuous colour of the shells, this species often escapes observation in places where it is really abundant, the least ripple on the surface of the water being sufficient to conceal it from an inexperienced observer.

It is sometimes found adherent to the stems and leaves of aquatic plants, and in such cases can be easily seen ; but as it prefers to fix itself to stones at the bottom of the river, and as the shell is nearly of the same colour as an ordinary stone, it looks more like an excrescence than a shell.

There is another species which is still smaller, but which prefers submerged plants to stones. It can readily be distinguished from the preceding molluscs by the position of the apex. In *Ancylus fluviatilis* the apex is directed to the right, while in *Ancylus oblongus*, as the second species is termed, it is directed towards the left.

At figs. 9 and 10 on the same plate are shown two remarkable objects, one of them looking something like a sponge, and the other resembling a slug covered with flowers. Both of these objects are allied to the molluscs, and belong to the curious group of animals called Polyzoa, because they consist of a number of separate animals massed together, and connected with each other. Although the sea is incomparably rich in Polyzoa, fresh water has many species, and two examples are here given as types of the whole group.

In lakes, ponds, and in still places in rivers, may be seen masses of a brownish-yellow hue, adhering to various substances, and especially to submerged wood. These masses have a great resemblance to sponge, and therefore go by the familiar name of Fresh-water Sponge. When dead and dry, they are even more sponge-like than when they are alive and full of moisture; but they really have no connection with the true sponges, being placed higher in the scale of creation than the crabs, insects, spiders, worms, &c. We will, however, use the popular word when talking of them. Several species of these curious beings are known in England, but we will content ourselves with that which has been mentioned.

The general appearance of the Fresh-water Sponge can be seen by reference to fig. 9, which represents a small specimen attached to a twig. The pendent twigs of trees which grow by the water-side, and allow their twigs to droop below the surface, are common resting-places of this creature, which sometimes covers all the twigs with its members, just as the oysters cluster on the submerged mangrove-boughs in the West Indies.

When it is undisturbed, its lower surface is seen to be covered with a delicate downy film of whitish colour; but if it should be touched, the film vanishes in a moment, leaving the lower surface exposed.

The apparent film is in reality an aggregation of the animals of which the sponge is composed, and if the aquarium-keeper can manage to place a little piece under the microscope while it is in full vigour, he will see a beautiful sight.

The whole of the surface is then seen to be thickly studded with little holes, from each of which projects a sort of gelatinous bell, with its edges covered with delicate tentacles, and resembling those which are drawn at fig. 10. These little objects are the inhabitants and architects of the sponge, and are closely allied to those which inhabit the Sea-mats, the Dead-man's Finger, and other marine polyzoa.

By means of the tentacles, the tiny animals arrest any floating particle which serves for food,

and although they seem to be devoid of all organs
of sense, they can take alarm at the approach of
any object, and in a moment all the pretty bells
with their curious heads will retreat into the
interior of the sponge, vanishing as if by magic.

Then, if the observer can manage to cut the
sponge in two directions, namely, at right angles to
the holes into which the animals have vanished,
and also parallel with them, he will see that the
whole of the apparent sponge is composed of tubes
lying by the side of each other, and forming places
of refuge into which the creatures can retreat.
Indeed, when placed under a microscope, they bear
a tolerably close resemblance to the aggregated
tubes of the Sabella, which has already been described.

The scientific name of this Fresh-water Sponge
is *Alcyonella fungosa*, the latter term being given
to it, because its brown masses might easily be
mistaken for submerged fungi.

At fig. 10 is seen a curious object hanging over
a stick, and having its back covered with flower-
like protuberances. This is the *Cristatella mucedo*,
one of the most singular beings that England, or,
indeed, any country, produces, and which should
be welcomed as a valuable and interesting adjunct
to the aquarium.

The example shown in the illustration is a rather
larger specimen than is generally found, the usual
length being about an inch. It looks much as if it
had been made of soft, woolly silk, and at first
sight bears a kind of resemblance to one of the
long-haired caterpillars that had been flattened by
accident.

The strange point in this creature is, that it is
locomotive, and can crawl along with a steady and
slow motion. How it manages to crawl is really a

wonder. If it were a single animal, there would
be no mystery about it; but the reader must
remember that it is an aggregation of separate
animals, each distinct in itself, though preserving
a connection with all the others.

To all appearances, these delicate bell-like animals
have no particular organs of sensation, and their
tentacles are merely outspread as nets wherein may
be entangled any particles of food that may happen
to float against them. But the animals cannot see
their food, neither can they direct their tentacles
towards it; and it seems almost incredible that
beings thus constituted should have any power of
concerted action. Yet, that such is the case is
evident from the fact that the aggregated mass is
able to crawl about ; for, unless the multitudinous
animals agreed to move consentaneously, the entire
mass would remain still.

Some persons have said that there is really no
concerted action, but that the movement is simply
involuntary, like the action of the heart in the
higher animals, and that it does not imply any
consciousness on the part of the Cristatella.

Now, if the movement were always in one
direction, or if the compound animal were to
wander about in a purposeless manner, there might
be some grounds for such a supposition. But, in
fact, the Cristatella does clearly choose the direction
in which it goes, inasmuch as it shuns the dark
and shady places, and loves to crawl in shallow
water, where it can enjoy the light and heat of the
sunbeams.

Whatever may be the impelling motive, the
organ of movement in the Cristatella is evident
enough. It consists of the base of the compound
animal, which is flattened, very contractile, and is
used much as the foot of a gasteropodous mollusc.

The Crustacea which inhabit the fresh water can but have a brief notice in these pages.

The largest and finest of the Fresh-water Crustacea—namely, the common Crayfish, cannot conveniently be kept in an aquarium, as it is too bulky for such limited space, and is, besides, so voracious that it would devour every other animal in the tank. It can be kept for some time in a separate vessel, but is always a troublesome creature to manage, as I can testify from much experience. As to the Shrimps and Prawns, several species of them make their way up tidal rivers, and penetrate so far inland that they only taste salt water at high tides. Any number of shrimps, prawns, and crabs can be taken in the Thames, near Woolwich; but still, they are not strictly denizens of the fresh water, being only stragglers from the sea.

There are, however, some fresh-water Crustacea which are small enough for the aquarium; and we will take as our typical example the common Fresh-water Shrimp, shown at Plate VII., figs. 7 and 8.

In general structure the Fresh-water Shrimp bears some resemblance to the common Sand-hopper, to which, indeed, it is closely allied, and its movements in water increase the similitude.

Fresh-water Shrimps may be found plentifully in almost every streamlet or ditch, provided that the water be tolerably clear. They act much like the fish in their habit of keeping their heads up the stream, and in their general conduct look something like the fry of various fish. Sometimes they make their way up the stream by clinging to the stones and other objects that form the bed of the stream, making quick darts forward, and then holding tightly to a stone until they choose to make a second dash onwards.

When they have gone up the stream as far as they think proper, they loosen their hold and come drifting back again, sometimes rolling over and over, but generally contriving to keep their heads pointing up the stream. In fact, they appear to amuse themselves by this action, just as the gnats amuse themselves by dancing up and down in the air.

The food of the Fresh-water Shrimp is usually decaying animal matter, and it can be attracted by sinking a piece of half-putrid flesh in the water. When it is not engaged in active exertion, it retires to some little crevice at the side of the stream, whence, however, it keeps a careful watch, so as to be able to dart out in a moment as soon as it sees anything floating past which looks as if it might be eatable.

When removed from the water, the little crustacean is quite helpless, lying on its side, and merely spinning round and round in its struggles, —a habit which has gained for it the title of Fresh-water Screw. Its scientific name is *Gammarus pulex.*

As for all the great tribes of the Entomostraca, I do not enter upon them here, as they are mostly exceedingly minute, and require too much special attention to be placed in an ordinary aquarium. Some specimens will most certainly be found therein, and if the young observer wishes to watch them, he should take them out and place them in a vessel by themselves. One or two species will be briefly mentioned in a future page.

The best way of removing them is by means of a piece of glass tube.

Take the tube and press the forefinger lightly upon the top, so as to exclude the air. Then push it into the water just above the animal to be captured. Remove the finger suddenly, and the water will rush up into the tube, carrying the creature with it. Replace the finger, and the animal can then be removed in the tube, and so transferred to the proper vessel.

On Plate VII., fig. 3, may be seen a very interesting creature, and one that is a great favourite in aquaria. This is the common Water Spider, known scientifically as *Argyroneta aquatica*.

The Water Spider is remarkable for its custom of spending the greater part of its time under the surface of the water, only ascending to the upper regions for the sake of catching prey, or obtaining a fresh supply of air for its subaquatic home. The manner in which it makes its nest and supplies it with air is singularly interesting.

Like all other similar creatures, it can exist for a considerable time without needing to breathe. It begins by crawling down the stem of some aquatic plant, there spinning a closely-woven and dish-like web. It then ascends to the surface, protruding the end of the abdomen, and, with a quick jerk, seizes a bubble of air and instantly dives. This bubble is partly retained by the thick coating of grey hair with which its body is covered, and partly by the hinder pair of legs, which are crossed, and serve to hold it in its place.

As the spider descends, the bubble shines as if it were polished silver, and the look of the creature is really pretty. As soon as it has reached its web, it crawls underneath it, uncrosses its legs, and thus releases the air-bubble, which is caught under the web and buoys it upward. The spider then enlarges the web by adding to its edges, and as fast as she enlarges it, she ascends to the surface and brings down fresh supplies of air.

In order to guide itself in its ascent and descent, it makes use of a thread, one end of which is attached to the nest, and the other to some object on the surface.

The completed nest or cell is about half the size of an acorn, and much of the same shape. Within it the Water Spider passes the chief portion of its life. Within it the eggs are laid, some hundreds in number, enclosed in a separate cell and fastened to the upper part of the nest. Within it the young are hatched, and swarm over it, a crowd of little black dots soon to expand into full-blown Water Spiders. Within it the Spider takes its meals. It leaves the nest in order to catch prey; but as soon as it has done so, it returns to the cell, and there consumes its unfortunate victim. While at rest the Water Spider always remains in its cell with the head downwards, after the manner familiarized to us by the common spider of our gardens.

In taking leave of this curious, interesting creature, I venture to offer a few words of advice. Catch it if possible, and keep it for the purpose of watching its proceedings. Supply it plentifully with food, and place it, as far as can be done, in the same condition which would have surrounded it in its normal state. But, after having carefully noticed all its doings, and so carried out the object for which it was caught, take it to some suitable piece of water and set it at liberty. If it is kept in an aquarium, it can scarcely live its full life, and cannot live a happy life. Shorten, therefore, the term of its captivity as much as possible, and as soon as its purpose is served, set it free.

CHAPTER XII.

INSECTS.

WE will complete this work by a brief notice of the few out of the many insects that are suitable for the fresh-water aquarium.

It must be well understood, however, that all these insects are not supposed to exist at one time in a single aquarium, for the very good reason that the number of inhabitants which a limited amount of water can properly hold is very small indeed, and that it is always better to understock ·the aquarium than to run the least risk of crowding it.

Our list begins with the Water-Beetles, of which the Great Water-Beetle (*Dyticus marginalis*) takes the first place. This fierce and voracious insect, which is shown at Plate X., fig. 7, is very plentiful in ponds and ditches, and may be found in all its stages of development.

The larva, which is shown at Plate IX., fig. 12, is a formidable-looking creature, attaining the length of two inches, and being proportionately stout. It is armed with a pair of sickle-like jaws, with which it captures other insects. These mandibles are hollow, like the fangs of venomous serpents, and have near their extremity a little aperture through which the insect sucks the juices of its prey. As soon as the larva seizes any unfortunate insect, it bends itself backwards so forcibly that the head nearly rests upon its back, and in this position it remains until it has finished its repast.

The reader will see that at the end of the tail there is a brush-like appendage. This structure is a promulgation of the breathing-vessels, and enables the larva to breathe atmospheric air by thrusting them above the surface. It often hangs suspended, as it were, in the water, the head being below, and the breathing appendage on the surface.

When it has finished its growth, it burrows into the earth and makes a rounded cocoon, in the interior of which it undergoes its transformation, first in the pupal form, and next in its perfect state.

In its beetle form it is quite as fierce, quite as voracious, and far more active than it was in its imperfect stages of existence. It can swim faster than it could when in the larval state, and can kill and eat insects and other creatures far larger than itself. Even the fishes are not safe from its powerful jaws, and the beetle has a habit of diving quietly under them and biting out a piece of flesh from the softer part of the abdomen.

Moreover, it is gifted with a pair of large and powerful wings, by means of which it can transport itself from one piece of water to another. Those, therefore, who keep this insect in the aquarium must be careful to cover the vessel, or the beetle will be sure to escape if the opening be left unprotected.

Foot of Dyticus.

The young naturalist should be careful to examine the fore legs of this beetle. If it should

happen to be a male one, the joints of the fore feet
will be seen to be expanded into a most curious
apparatus of suckers. Even to the naked eye,
this structure is a very interesting one; but its
full beauty cannot be appreciated until it is examined by means of a microscope. The best mode
of seeing it properly is to prepare one specimen
with Canada balsam, after the usual manner, and
to keep a second specimen which can be viewed as
an opaque object.

Only one of these beetles can be kept in a single
vessel, or it will kill and eat every other inhabitant
of the aquarium; and if a second specimen be
placed in the same vessel, they will fight until the
weaker is killed and eaten by the stronger.

Another common water-beetle is that which is
known by the name of *Acilius sulcatus*, a figure of
which may be seen in Plate X. The larva of this
insect is shown in Plate IX., fig. 7, and, as may be
seen, bears a considerable resemblance to that of
the Dyticus. It may, however, be distinguished,
not only by its smaller size, but by the great
length of the first segment of the body, which is so
elongated as to look like a neck. The body, too,
is much smoother than that of the Dyticus.

The long, slender body is very flexible, so as to
give it a very snake-like aspect, which is strengthened
by the mode in which it takes its prey. When it
wishes to feed, it swims very quietly under its
victim, and then grasps it firmly in the sharp jaws,
and drags it under water.

In its perfect state it is as active as the Dyticus,
and has the power of producing a loud humming
sound. It can fly as well as that insect; and if it
happens to fall on its back, has a way of regaining
its feet by leaping into the air, just as is done by
the well-known Skip-jack Beetles.

On Plate IX., fig. 4, may be seen the larva of an allied genus called Colymbetes ; and the perfect insect may be seen in Plate X., fig. 3. This beetle is very strong on the wing, and has been known to fly its rounds at night, attracted by the light of a lamp.

All these beetles, although mostly found in the water, are quite independent of that element, and are forced to breathe atmospheric air. They procure the air by rising to the surface of the water and protruding the end of the body. A quantity of air is then taken into the space between the body and the elytra, or wing-cases, and on this supply the insect is able to support respiration for a considerable time, when it is again obliged to return to the surface and take in a fresh supply of air.

On account of the necessity for respiration, this insect can easily be discovered, as the observer has only to watch by a pool or ditch for a few minutes, when if any of these beetles inhabit tho water, some of them are sure to come to the surface, rapidly take in the needful air, and dive again to the bottom.

At fig. 5 on Plate X. may be seen a figure of a little beetle which is popularly known as the Whirlwig, and scientifically as *Gyrinus*, both names having a similar signification.

There are several species of this genus, and they may be at once recognized by their habit of swimming about on the surface of the water, whirling about in their mazy dance with wonderful swiftness; and being very fond of company, it is very seldom that a solitary Whirlwig Beetle is found, and as a general rule they are gathered together in little companies of twenty or thirty in number.

Like the other water-beetles which have been mentioned, the Whirlwig has a pair of large wings,

by means of which it can fly to considerable distances. The chief use of these wings is to transport
it from place to place, in case the pond or puddle
in which it lived should be dried up by the summer
heats.

The Whirlwig is a good insect for an aquarium,
if it can only be kept alive, as it keeps up its merry
gyrations throughout the year, and even in the
winter time little companies of these pretty beetles
may be seen in sheltered spots whirling about as
briskly as in the middle of summer. When Whirl-
wigs are placed in an aquarium, the observer should
be careful to look at the eyes, which are so formed
that the insect can with equal ease see objects
below and above it.

The very remarkable larva of the Whirlwig is
shown on Plate IX., fig. 11 ; the figure being, of
course, much enlarged. The chief peculiarity in
its form is to be found in the numerous projections
from the sides. These projections are transparent,
and are connected with the organs of respiration, a
delicate air-vessel traversing each of them, and
being connected with the larger vessels within the
body.

As soon as the larva has completed its growth,
it crawls up the stem of some aquatic plant, and
spins a greyish cocoon, in which it remains until it
has attained the perfect state.

There is a large beetle much resembling in shape
and general appearance the Dyticus, but which
attains a greater size, and may be recognized by its
black colour, and the coating of silvery-white down
upon its under surface. This is called *Hydrophilus
piceus*, and is shown at fig. 2 on Plate X.

The larva is drawn on Plate IX., fig. 8, and is
much like that of Dyticus, but it may easily be
distinguished by the appendages to the abdomen,

which are short and threadlike instead of being
broad and hairy. The upper side of the head too
is flat, and the under convex. The jaws are not
as sickle-shaped as those of the Dyticus, and are
each armed with a tooth near the middle. Its
grub grows to a considerable size, often reaching
three inches in length, and being more stout in
proportion than that of the Dyticus. The larva
attains its full growth towards the end of summer,
when it leaves the water, and forms a burrow in
the earth, in the end of which it forms an oval
cocoon. In spite of its large size, it does not
occupy quite four months in passing through the
whole of its changes, and, according to Mr. West-
wood, it spends about sixty days in its larval state,
and thirty in the cocoon.

If possible, the aquarium-keeper should procure
a female Hydrophilus, for the sake of watching the
remarkable manner in which it lays its eggs. Instead of depositing them singly, it arranges them,
some sixty or so in number, symmetrically together,
and spins around them a kind of cocoon, formed of
fine white silk within, and covered with a gummy
secretion, which is impervious to water.

The cocoon is shaped somewhat like a small
turnip, and has a sharp point, which coincides with
the root of that vegetable. It is fastened to some
water-plant, and when the eggs are hatched the
little grubs escape through the bottom of the cocoon.
About five or six weeks generally elapse before
the eggs are hatched, and the larvæ change their
skins three times before assuming the perfect condition.

This is a very excellent insect for the aquarium,
as it is neither so fierce nor so voracious as the Dyticus, feeding chiefly upon aquatic plants, and only
eating animal food at intervals. Consequently,

there is little danger in keeping several specimens of Hydrophilus in an aquarium, though if a single Dyticus be placed among them, it will certainly attack and kill them.

The little figure on Plate X., namely fig. 6, represents the last water-beetle which will be mentioned in this work. Its name is *Helophorus aquaticus,* and it scarcely looks like a water-beetle at all, not having the peculiar swimming legs of those species which have been described. In fact, it crawls rather than swims, preferring to creep on the stems and roots of aquatic plants to swimming freely in the water. It walks very slowly, and occasionally leaves the water and crawls on the banks, in which case it has not a very prepossessing appearance, being covered with the mud in which it has been traversing. There are many species of these smaller water-beetles, but those which have been described will serve as types for almost any species that are likely to be captured.

The whole of Plate XI. is occupied with a group of insects which are tolerably familiar to those who live in the country, and have been in the habit of watching the various beings that inhabit the water. Figs. 2 and 11 exhibit two specimens of the insects known popularly as the Caddis-flies, while the remainder, from figs. 3 to 12 inclusive, depict the insect in its various stages, and the curious habitation which it constructs.

All anglers are well acquainted with the Caddis-worm, which is so useful a bait, and which will often entice a refractory fish which cannot be induced to yield to the temptation of an ordinary bait. These so-called " worms " are the larvæ of various Caddis-flies, about a hundred and eighty species of which are known to inhabit this country. It is evidently impossible to describe all these species, and I will therefore select two as examples.

Fig. 11 shows one of the largest species, *Phryganea grandis*, as it appears when in the act of flying; and fig. 2 shows a smaller species, *Leptocerus niger*, in the attitude which it assumes when walking. These insects can run with some speed ; and even if they should fall into the water, they can run along its surface with considerable rapidity, leaving a long wake behind them, and being generally snapped up by some hungry fish before they can reach the bank. The last-mentioned insect is notable for the great length of the slender antennæ, which remind the entomologist of the same organs in the Japan moth.

The eggs of the smaller insect are deposited beneath the water, being attached in little bundles to the stems of aquatic plants, and there left to be hatched. When the young larvæ pass from the egg into the water, they begin to form for them- selves habitations of a rather remarkable character.

Their bodies are white and soft, and would attract the notice of any fish that might pass within a reasonable distance of them. Taught by a wonderful instinct, they gather various substances, and fasten them together so as to form a tubular house, in which they live.

When they wish to move about, they protrude their black horny head from one end of their habitation, as well as the little legs which are attached to the first few segments of the body, and crawl slowly along the plants or the stones and sticks at the bottom of the stream. While in their homes, they bid defiance to the fish, for as soon as they are alarmed, they withdraw themselves entirely within their cases, and lie concealed until the danger is past.

Their habitations are made of all kinds of materials, depending partly on the particular species, and partly on the locality in which they have been hatched.

Several varieties of these remarkable habitations are given in Plate XI. Fig. 3 is one which has been made of sand and small stones. Sometimes these sand houses are curved, and look something like tiny horses.- Fig. 4 is one which is formed of little scraps of grass and dead leaves: those of the fir are sometimes used in this manner. They are arranged in regular rows, and the whole structure is attached to a stick, which acts equally as a support and a balance. Fig. 5 exhibits a case made of the same materials, but constructed by a different species of Caddis, and which is made of larger strips fastened longitudinally together.

CHAPTER XIII.

INSECTS (*continued*).

IN the accompanying illustration may be seen specimens of cases which have been selected from a large collection. Fig. *a* is made of little bits of stone, while fig. *c* exhibits two specimens of the curved sand cases which have already been mentioned. Fig. *e* is a peculiarly good specimen of a case which is formed of bits of grass - stems cut to measure, and ingeniously set spirally, so as to leave a tubular aperture in the centre, in which the little architect lives.

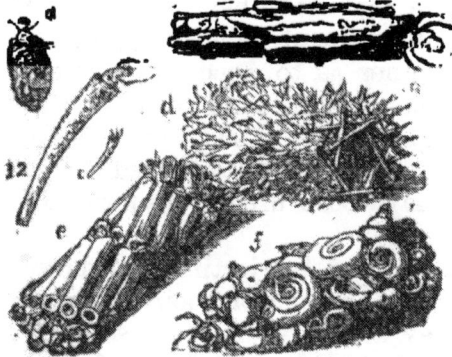

At fig. *b* is another case, made of fragments of decayed wood; and at *d* is a singularly ingenious one, made of scraps of grass arranged crosswise over each other. When first found, the grass was stil, green, but in process of time it became dry withered, and almost colourless. A somewhat similar specimen is seen at fig. 5 on Plate XI. but the grass is mixed up with other materials, and is not so regularly arranged.

At fig. *f* is an example of a kind of Caddis-case

that is often found, and which is composed of small shells, mostly belonging to the genus Planorbis. Two other examples may be seen on Plate XI., figs. 6 and 7. In making these cases, the Caddis-worm is by no means particular as to the shells, and is just as likely to use those which have living occupants as those out of which the little molluscs have died.

The young aquarium-keeper is strongly advised to procure some Caddis-worms, and to watch them at work. They are most interesting creatures, and well repay the trouble of examination. If they are quietly ejected from their homes, and supplied with fresh materials, they will soon set to work to build fresh habitations, and by a judicious regulation of the supply of material, they can be forced to build cases of all sorts of substances, and to do so in such a manner as to produce a kind of pattern. Brilliant substances, such as fragments of coral, coloured glass, and sands, china, shells, straws, and so forth, have been supplied to the Caddis-worms, and have been used in the formation of their new homes.

Great care must be taken in ejecting these larvæ, as they have pincers at the end of their bodies, by means of which they attach themselves so firmly to their homes, that if roughly pulled out of them, death is very likely to be the result. The entomological reader will see a curious analogy between the habit of this larva and that of the common clothes-moth, each making itself a tubular home, and each being capable of forming a structure of materials artificially supplied to it.

Within these curious habitations the grub passes all its wingless life. As soon as it has attained its full growth, it closes the end of the case with a strong silken network, which permits the water

to pass through it, but is an effectual barrier
against foes. It then becomes a pupa, and remains
in that state until the time comes for it to assume
the perfect form. The pupa is shown on Plate XI.,
fig. 10.

On the same plate, figs. 13 and 14, are seen two
odd-looking creatures which can be found easily
enough in our streams by passing a net along the
banks, and poking up the mud with a stick. These
are preliminary forms of two species of those beautiful insects which are so well known under the
general title of Dragon-flies. In some parts of the
country they are called Horse-stingers; but it is
scarcely necessary to remark that they have no
stings, and are perfectly harmless.

In their arval and pupal states, these insects are
very interesting inhabitants of the aquarium; but
as they are fierce and voracious, they must not be
placed in the same vessel with other creatures.

Fig. 14, Plate XI., shows the larval state of one
of our common Dragon-flies, known scientifically
as *Libellula depressa.* It may be easily known by
its rather short and flattened body. If the reader
will look at the extremity of the tail, he will see
that it is terminated by several pointed projections.
Altogether, there are five of these projections, three
being much larger than the others. They can be
separated or brought together at the will of the
insect.

When these plates are separated, a passage is
opened into the interior of the body, into which
the water can pass, and with which the respiratory
organs are connected. As soon as the insect has
extracted the oxygen, the water is expelled from
the opening into which it was received. As this
expulsion can be accomplished with considerable
force, the insect is driven forward by the reaction

of the column of water which is ejected, and is propelled in the same way as the various cuttle-fishes. · Engineers have now taken up this principle, and have built engines which propel ships by taking in a quantity of water slowly, and expelling it violently through a tube.

The accompanying illustration exhibits the larva of another species of Dragon-fly; one of the pretty demoiselles that flit about the water, and whose green or blue wings flash so splendidly in the sunbeams. In this species the comparatively large size of three of the projections is very plainly shown. The perfect form of this insect is shown at fig. 13, Plate XII.

This is an insect nearly full of wonders. It has a mode of propelling itself almost unique, and it has a mode of seizing its prey which is equally surprising.

Larva of Dragon-fly.

If the head of the larva be carefully examined, it will be seen to be furnished with a remarkable modification of the jaws. The lower lip is made of four pieces—the two first being used as pincers, something like those of the crustaceans, and the two others forming a jointed apparatus by which the pincers can be laid close to the face, or extended, at the wish of the insect.

A tolerable notion of this apparatus may be gained by pressing the hand upon the mouth and keeping the arm close to the body. The hand now represents the pincers and the arm the two joints of the lip. By means of this curious structure the Dragon-fly larva can suddenly dart out the pincers, seize a passing insect with them, and hold it against its mouth while the cruel jaws devour it. The food of this voracious larva consists of any small living creature that may come within its reach, and even the young of various fishes are not spared. The details of this apparatus, which is technically called the "mask," vary in different species, but the principle of the structure is the same in all.

The mode in which these insects escape from the pupal condition, and attain their wings, is singularly interesting, and ought to be watched by the aquarium-owner. In these creatures the change from the larval to the pupal form is not so pronounced as in the generality of the insect tribe, the pupa resembling the larva in shape, and being quite as voracious in the second stage as in the former.

At last, however, it ceases to feed, and becomes dull and languid, looking as if it were going to die. However, it has only felt that it has outgrown its imperfect condition, and is about to exchange its dull existence beneath the waters for a brilliant aërial life—a life not less active nor less ferocious than that which it formerly spent beneath the water. It crawls slowly and languidly up the stem of some aquatic plant, and when it has reached the height of a foot or so above the surface, it pauses and awaits the change. Presently the back begins to split, and after a few convulsive struggles the Dragon-fly breaks out of the pupal shell in which it has been hidden.

For a while it remains in the same spot, shaking out its beautiful wings, which at first are mere

shapeless projections, but which soon gain their
proper form and firmness in the sunbeams; and in
an hour or two the Dragon-fly takes flight, leaving
the empty pupal skin still clinging to the plant up
which it had crawled.

The appearance of the insect while employed in
the act of quitting the shell is shown at Plate XII.,
fig. 14; and in the summer time plenty of the discarded skins may be seen adhering to the water-
plants.

On Plate XII., fig. 1, may be seen a figure of
the common May-fly (*Ephemera vulgata*). Although
the short life of the perfect insect renders it useless
as an inhabitant of the aquarium, it is an interesting creature in the imperfect stages of existence,
and should be kept if only for the sake of the
singular manner in which it assumes its perfect
form.

Brief as is its life in the winged state, the Ephemera lives for three full years in the condition of
larva and pupa. The larva of the Ephemera
burrows in the muddy banks, making holes that
are shaped something like the letter U; so that
the creature can enter its burrow by one hole, and
leave it by the other, without having the trouble
of turning itself in its narrow domicile. The food
of the larva consists of vegetable substances, and,
from all appearances, the creature seems to swallow
much of the earth which is excavated in making
its burrow. As may be seen by the drawing on
Plate XII., fig. 2, the larva bears sufficient resemblance to the perfect insect to be recognized
without difficulty, and has the conspicuous filaments
at the end of the tail, though they are not so long
as in the perfect insect.

The most extraordinary part of the history of
this insect is its power of flight before it has entirely

thrown off the pupal envelopments. Like the pupa of the Dragon-fly, that of the Ephemera crawls out of the water, bursts from its pupa-case, and in due time takes to flight. But after it has flown about for a short space, it settles again, and then undergoes another change. The skin again bursts, and from its delicate envelope there issues a far more perfect form of insect. The wings are more delicate and gauzy, and the filaments of the tail increase to twice their original length.

The cast skin is left adhering to the object on which the yet imperfect insect had settled, and hundreds of these discarded envelopes may be seen upon the limbs of trees near water which is inhabited by the Ephemera. Their flight before casting off this pellicle is clumsy and constrained, and it is not until they are freed from its trammels that they are able to flutter up and down in the manner which is so characteristic of these insects.

We now come to a group of insects, many of which inhabit the water, not only in their larval and pupal stages, but even in their perfect forms.

The first of these creatures are those curious insects which are popularly known as Water Boatmen, and scientifically as *Notonectidæ*. Their name is derived from the boat-like form of their bodies, and the peculiar manner in which their last pair of legs are modified into oars. By means of these swimming legs, the Water Boatman rows itself about with much speed.

They are predaceous insects, chasing other inhabitants of the water, and killing them by means of a sharp beak-like appendage to the mouth. In some of the larger species this beak is sharp and strong enough to inflict a rather painful wound

upon a human being who handles them incautiously.
These insects have an odd habit of swimming on
their backs, a custom which has earned for them
the name of Notonectids, or Back-swimmers. One
of the larger species, *Notonecta furcata*, is shown
on Plate XII., fig. 3.

Another remarkable insect belonging to this
group is the Water Scorpion (*Nepa cinerea*).
Although it has no sting, and belongs to a totally
different order of beings, the Nepa does look very
much like a scorpion, the peculiarly formed prehensile legs and the pointed appendage to the tail
adding to the resemblance.

As in the case with the Water Boatman, the
Water Scorpion can fly well, its wings being large
and powerful. A figure of this insect as it appears
when swimming is shown
at Plate XII., fig. 4, and
the accompanying illus-
tration shows the same
insect with its wings ex-
panded as if in the act of
flying.

This is a very voracious
insect, and one that lives
entirely upon other inha-
bitants of the water.

Water Scorpion flying.

There is a very curious insect which is closely
allied to the Water Scorpion, but which is not so
plentiful. This is the Ranatra, a figure of which
is given on Plate XI., fig. 9.

This very remarkable insect is found in our
ditches and ponds, and is notable for its fierceness,
voracity, and indomitable courage. There is now
before me a fine specimen which was kept alive for
a considerable time by a young lady, and which
seemed to be quite as much at home in an aquarium

as if it had been in the Kentish streamlet from which it was taken.

As its captor did not know its real name, and did not wish to give it a title that had any appearance of scientific assumption, she called the creature "Daddy," because it looked something like a Daddy-longlegs. Daddy soon became an object of interest to the household, who used to amuse themselves by placing it in a small vessel together with various aquatic creatures, and watching its conduct. There were very few which it did not eat, but its favourites were the larvæ of Ephemeræ, and the fry of fishes. Even little shrimps fell victims to its voracity, which seemed insatiable.

The mode of capturing prey was very curious. It would crawl gently towards its victim, lift up its fore pair of legs, as seen at fig. 9, and strike sharply at the doomed creature. In a moment "Daddy" had its victim pressed firmly against its mouth, the fore legs being forcibly drawn back, and the short, sharp proboscis buried up to the base. At fig. 1, Plate XI., the Ranatra is depicted in the act of seizing its prey.

The rapidity and force with which it could strike were really astonishing, when the size of the creature was taken into consideration, and in proportion to its relative dimensions were far greater than can be found even in the lion or tiger.

"Daddy's" boldness was not less remarkable than its voracity. Though it disliked being disturbed, and resented removal from the aquarium in which it lived, it was only angered, and not frightened. If placed in a saucer with some water, it crawled about quite at its ease ; and if a little fish or a May-fly larva were introduced, it would instantly dart at it, enclose it in its deadly grasp, and hold it firmly until all its juices had been extracted. So totally devoid of fear was it, that if a finger were pointed at it, "Daddy" would raise itself indignantly, and strike as fiercely at the offending finger as if it had been a fish or a grub on which it was about to feed. "Daddy" lived for a considerable time in captivity; and when it died, it was placed in spirits and sent to me.

At fig. 8, Plate XII., is shown one of the curious insects which move so easily on the surface of the water, and are therefore called Water-Measurers. They use their middle feet as oars, by means of which they propel themselves, while the last pair of legs are trailed behind, and used as a rudder. This, like the other insects, has wings, and will therefore escape unless the aquarium be covered.

On Plate XII., at fig. 7, is a representation of the common Gnat (*Culex pipiens*), an insect which, though extremely disagreeable, is nevertheless extremely interesting, and is well worthy of being carefully watched. If possible, the mother insect should be procured in spring, in order that the remarkable method of depositing the eggs may be watched.

As the eggs are too heavy to float in water, and as it is necessary for their well-being that they should be exposed both to the air and the sun, the Gnat has the power of fastening them together so as to make a miniature boat, the general appearance of which may be seen by the accompanying illustration.

Eggs of Gnat. In fixing the eggs together, the Gnat makes use of the hind legs, which are crossed, and form a guide for the eggs.

This little boat is so beautifully made that it cannot be upset, and never sinks. Water may be poured over it, but not a drop enters ; and if pressed beneath the surface, it instantly floats again, always keeping the same end of the egg upwards. The upright position is necessary for the eggs, because the young escape into the water through the submerged end of the egg, which opens to permit their exit.

The larvæ are curious little creatures, with very large heads, slender bodies, and having a remarkable apparatus connected with the organs of respiration. They wriggle about in the water with great activity, assembling in great numbers at the surface, but diving at the least alarm.

If the reader will refer to the illustration of the larva on Plate XII., fig. 7, he will see that the end of the tail is furnished with a divergent tube, at the extremity of which is a star-like radiation of bristles. Through the centre of this tube air passes to the respiratory organs, and when the creature wishes to be at rest, it ascends to the surface, and lies with its head downwards, and the end of the tail-tube just projecting out of the water.

In process of time, it changes to a pupa, which is nearly as active as the larva. The organs of respiration are, however, arranged differently, being placed in two horn-like tubes which proceed from the thorax, and project above the water when the creature is at rest.

After passing a few days in the pupal form, it ascends to the surface of the water, and there waits until the skin bursts along the back. From the aperture the Gnat emerges, and dries its wings as it stands upon the shed skin, which serves as a raft.

Any number of Gnats, their eggs, larvæ, and pupæ, can be obtained from a rain-water tub which has been exposed to the air.

At fig 6, Plate XII., is shown one of our

handsomest flies, *Stratiomys chameleon.* It is
placed among the inhabitants of the aquarium on
account of its curious larva, which is seen at the
upper fig. 6 of the same plate. This curious larva
has its respiratory organs arranged something like
those of the common Gnat. From the end of the
tail proceeds a tube, which is terminated by a
radiating circle of hairs, in the centre of which is
the aperture leading to the organs of respiration.
When the creature is at rest, the end of the tube
just projects out of the water, the radiating hairs
being spread on the surface, and seeming, together
with the bubbles of air which they enclose, to act
as a buoy by which the weight of the body is supported. When it changes into the pupal condition,
it does not throw off the larval skin, but remains
within it; and as it is very much smaller in the
pupal than the larval condition, it only occupies a
comparatively small portion of the shed envelope.

The last insect to which we shall refer in this
work is the common Drone-fly (*Eristalis tenax*),
which is shown on Plate XII., at the lower fig. 6.
The perfect insect is very well known to all lovers
of the garden, as it flies from one place to another
with its peculiarly darting swiftness, and settles
upon leaves and walls, with its nimble body palpi-
tating as it respires.

The larva of this insect is a very singular-looking
creature, and is shown at the upper fig. 5,
Plate XII.

It lives only in stagnant and muddy waters, and
in old, neglected pools may be found plentifully.
The largest assemblage of these creatures that I
ever saw was in Wiltshire. A tub had been sunk
in the ground for the reception of water, and had
gradually become half filled with dead leaves and

other *débris*, which decomposed into a soft mud. This mud was so closely packed with the larvæ of the Drone-fly, that the water was quite choked with them.

This larva is sometimes called the Rat-tailed Maggot, on account of its peculiar structure. Like that of the preceding insect, its respiratory organs pass through the end of the abdomen ; but in this creature, the tube through which the organs of respiration communicate with the air is very long, and capable of being projected or withdrawn at pleasure. Even with the unaided eye, the tracheal tubes can be seen within this curious structure, while an ordinary pocket lens exhibits them admirably.

As the tiny Entomostraca have been casually mentioned, three figures are given on Plate XII., by which the reader may recognize the typical forms of these creatures.

At fig. 10 is seen an example of the Water Flea (*Daphnia*), the figure being, of course, very much enlarged. The Daphnias are light-loving creatures, and fond of congregating near the surface of the water, where they will sometimes assemble in such multitudes that they look like a muddy belt.

As the shell of the Daphnia is transparent, the body can be plainly seen through it ; and in the figure the eggs are seen, as they are kept between the shell and the body of the parent.

Fig. 9 on the same plate represents a very pretty species belonging to the genus Cypris. The shell of this creature is shaped very much like that of a bivalve mollusc, and completely encloses the body, only having an aperture through which the limbs can pass.

Figs. 11 and 12 represent the common Cyclops

in two positions ; fig. 11 showing the back, and fig. 12 the side. This little creature is very plentiful in our ponds and ditches, and may be easily procured for the aquarium. As the Entomostraca are so small, it is better to reserve a special vessel for them, placing in it a little duckweed, and allowing half an inch or so of mud at the bottom of the receptacle. A wide-mouthed bottle answers very well for this purpose, and nothing can be better than a confectioner's show-glass.

FINIS.

WYMAN AND SONS, PRINTERS, GREAT QUEEN STREET, W.C.

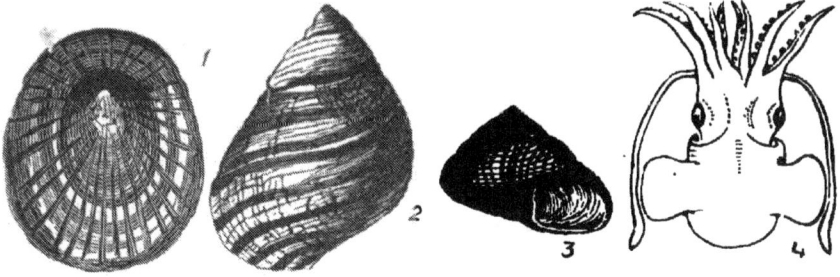

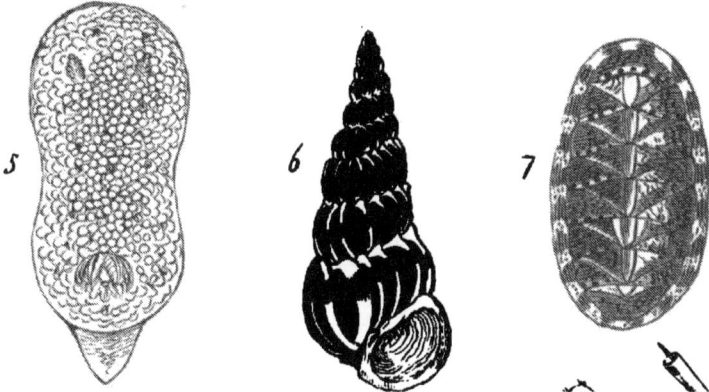

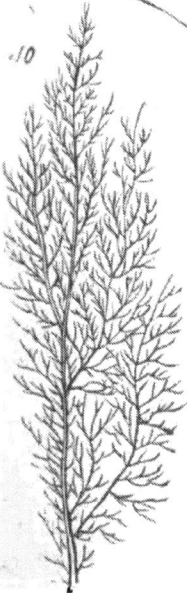

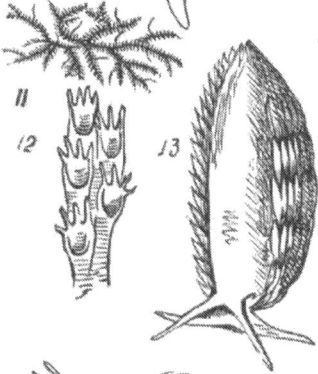

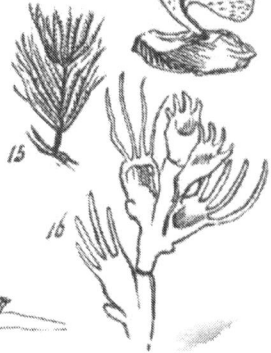

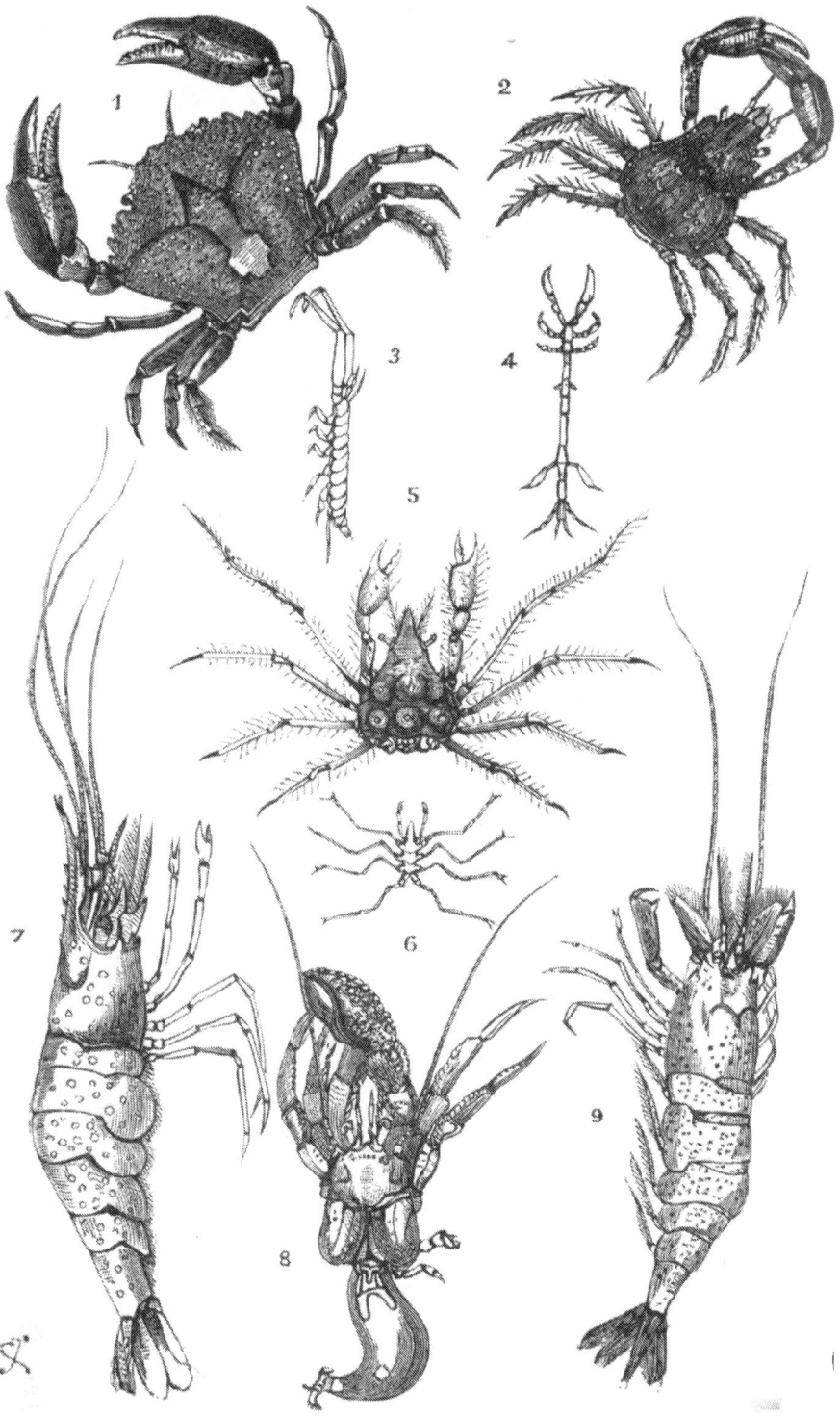

PLATE IV

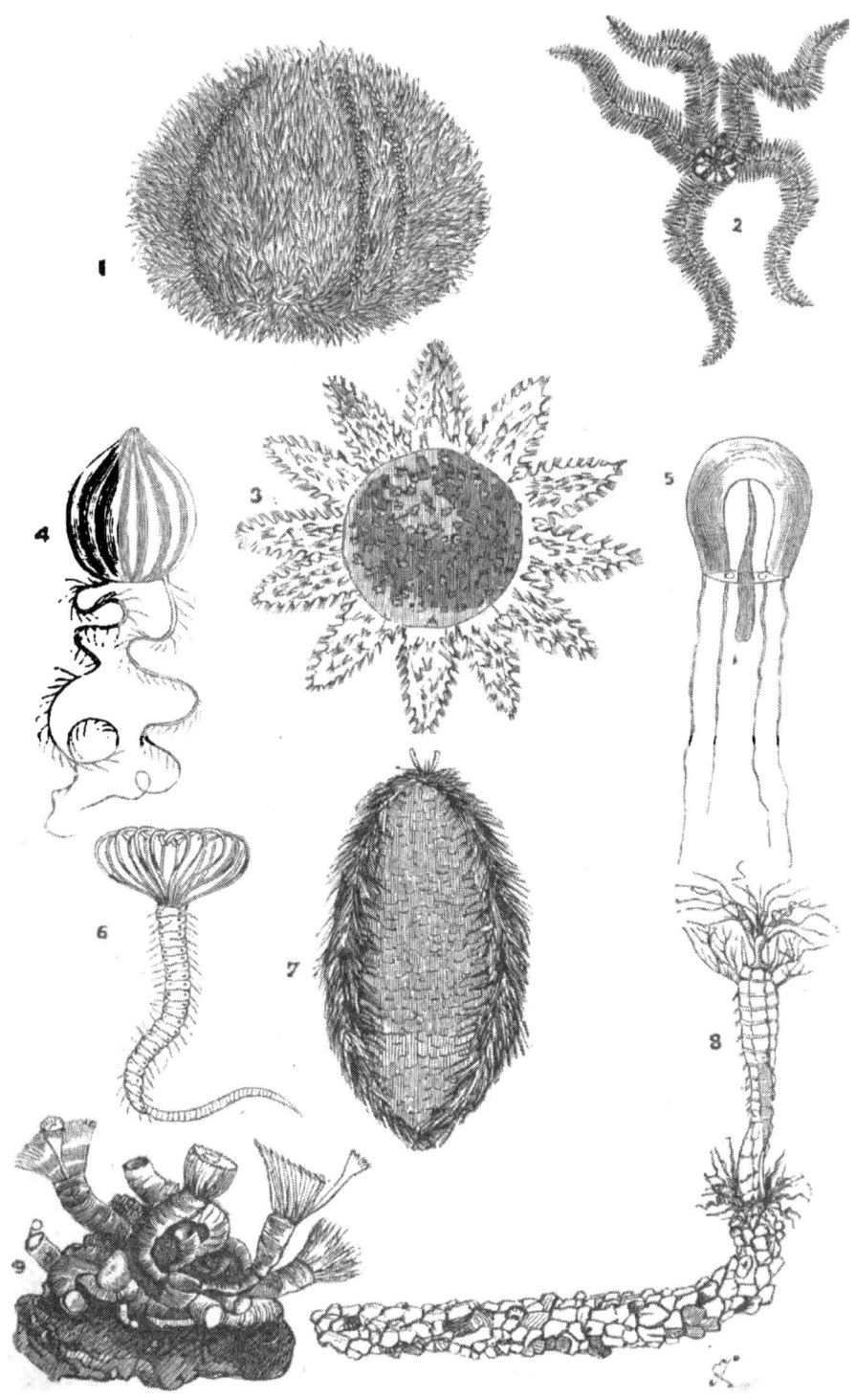

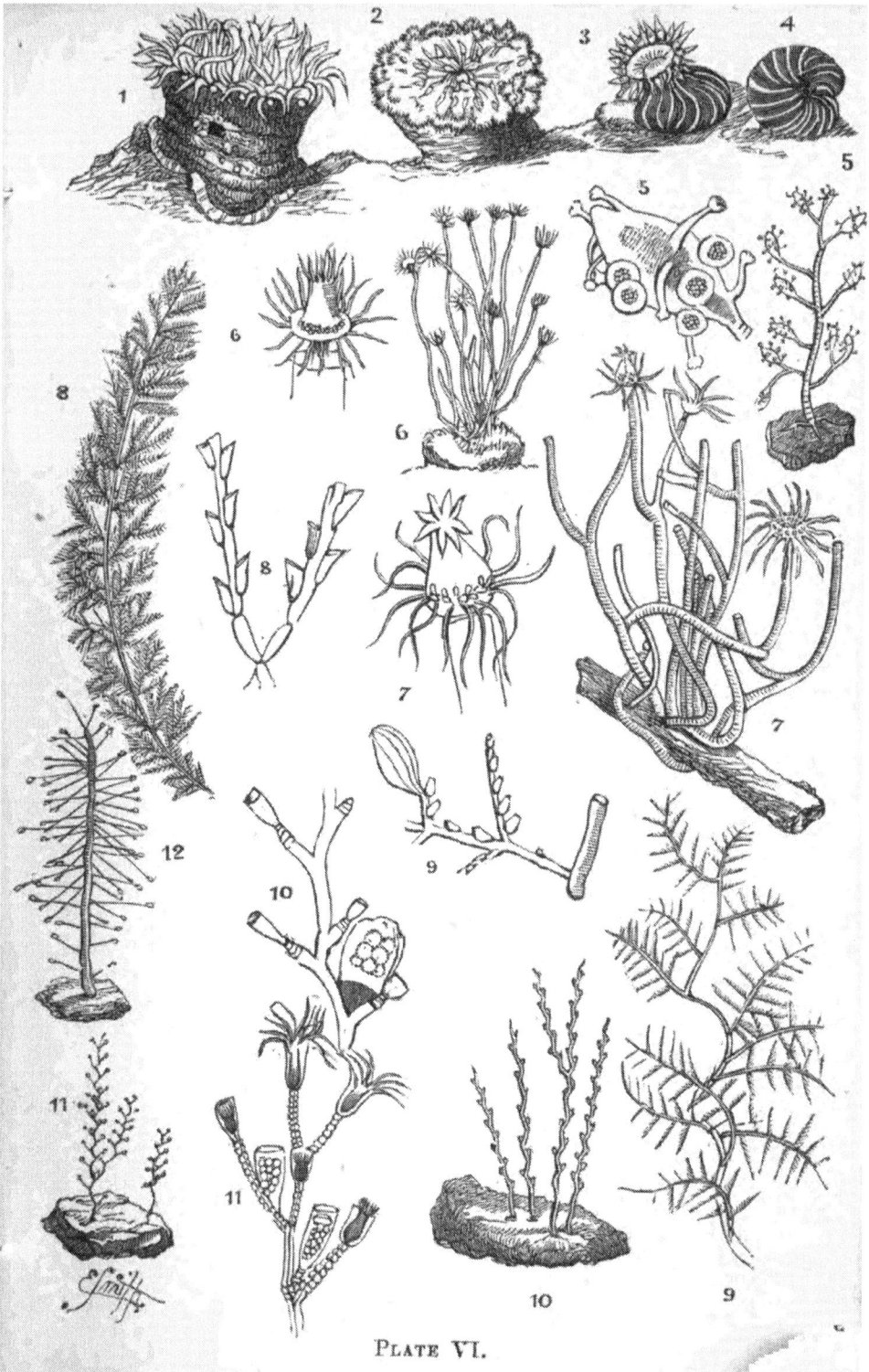

PLATE VI.

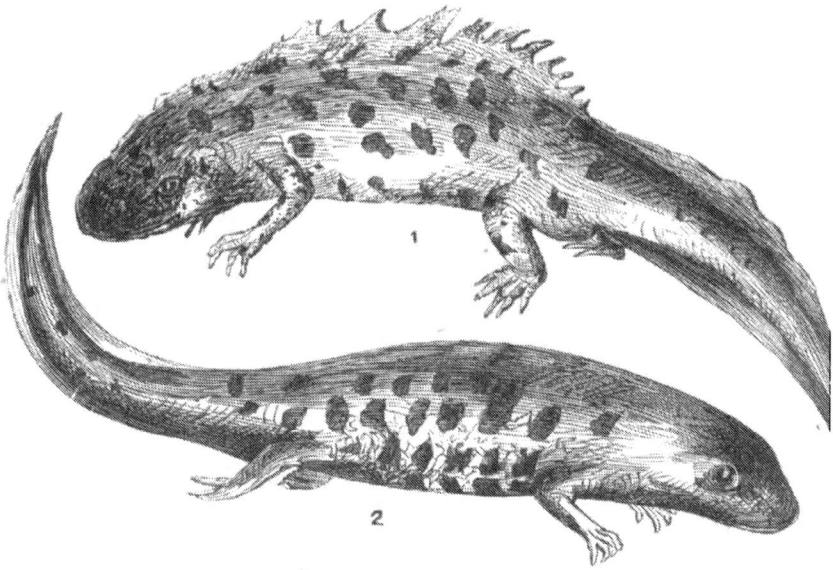

PLATE VII.

PLATE VIII.

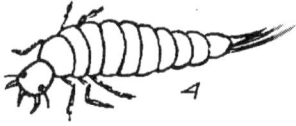

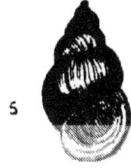

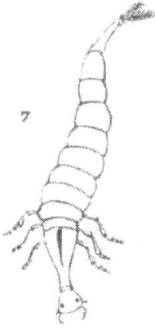

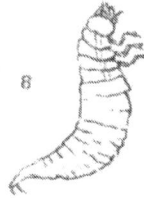

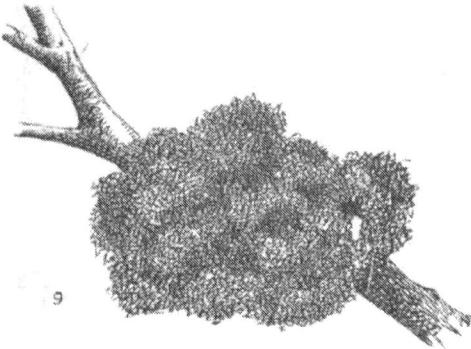

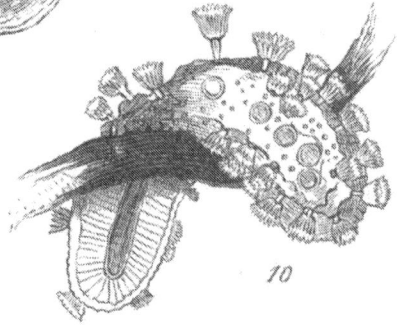

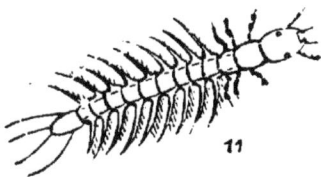

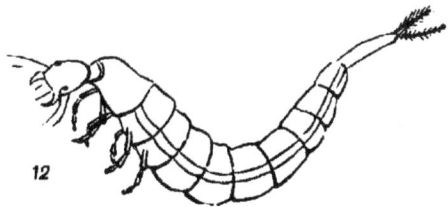

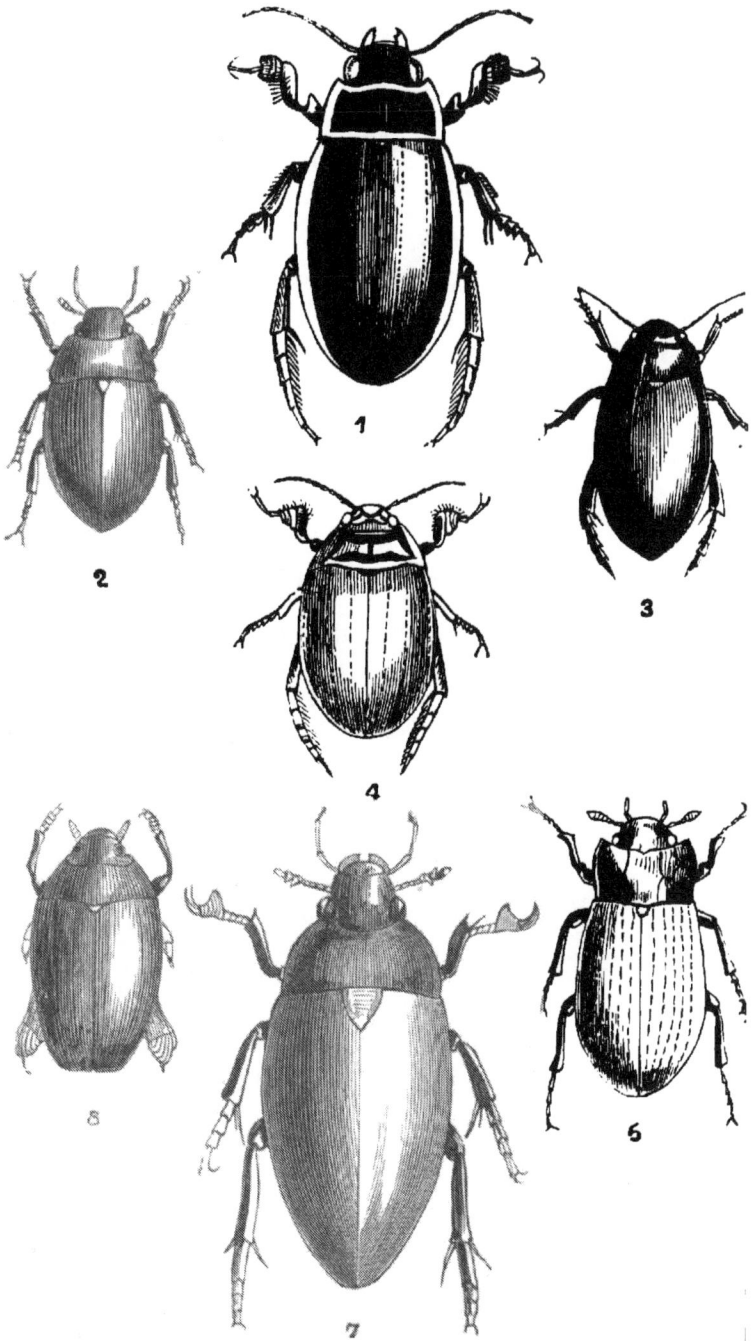

PLATE X.

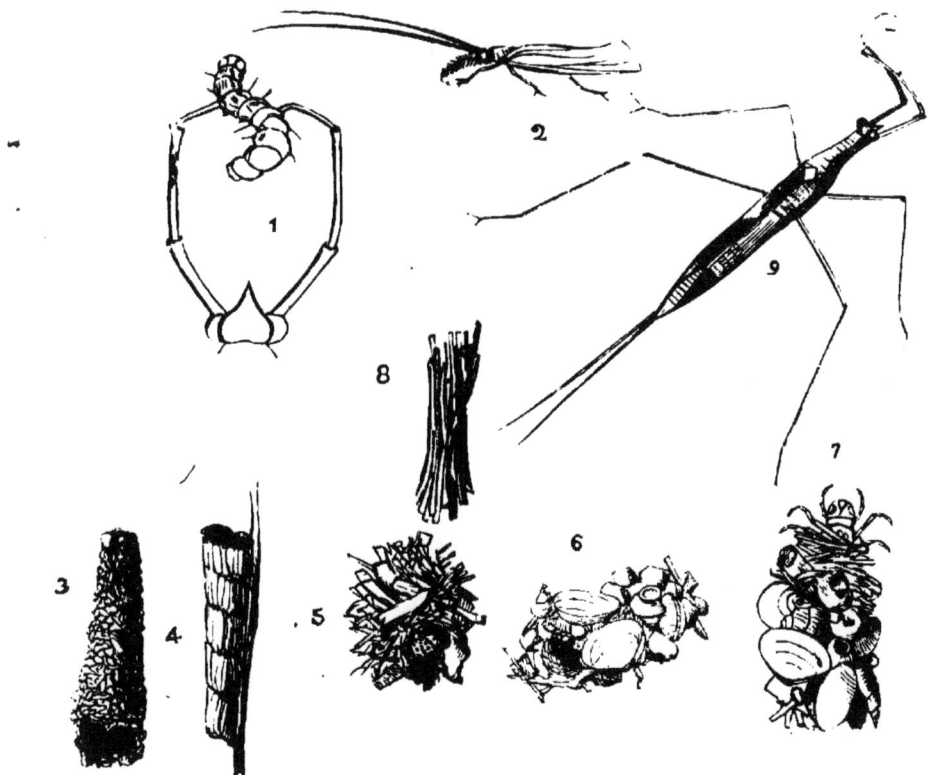

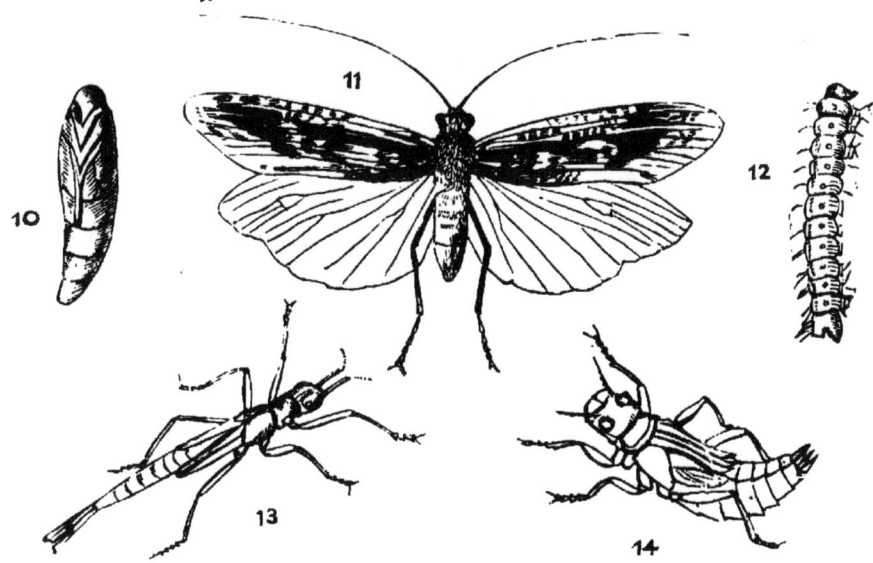

PLATE VI

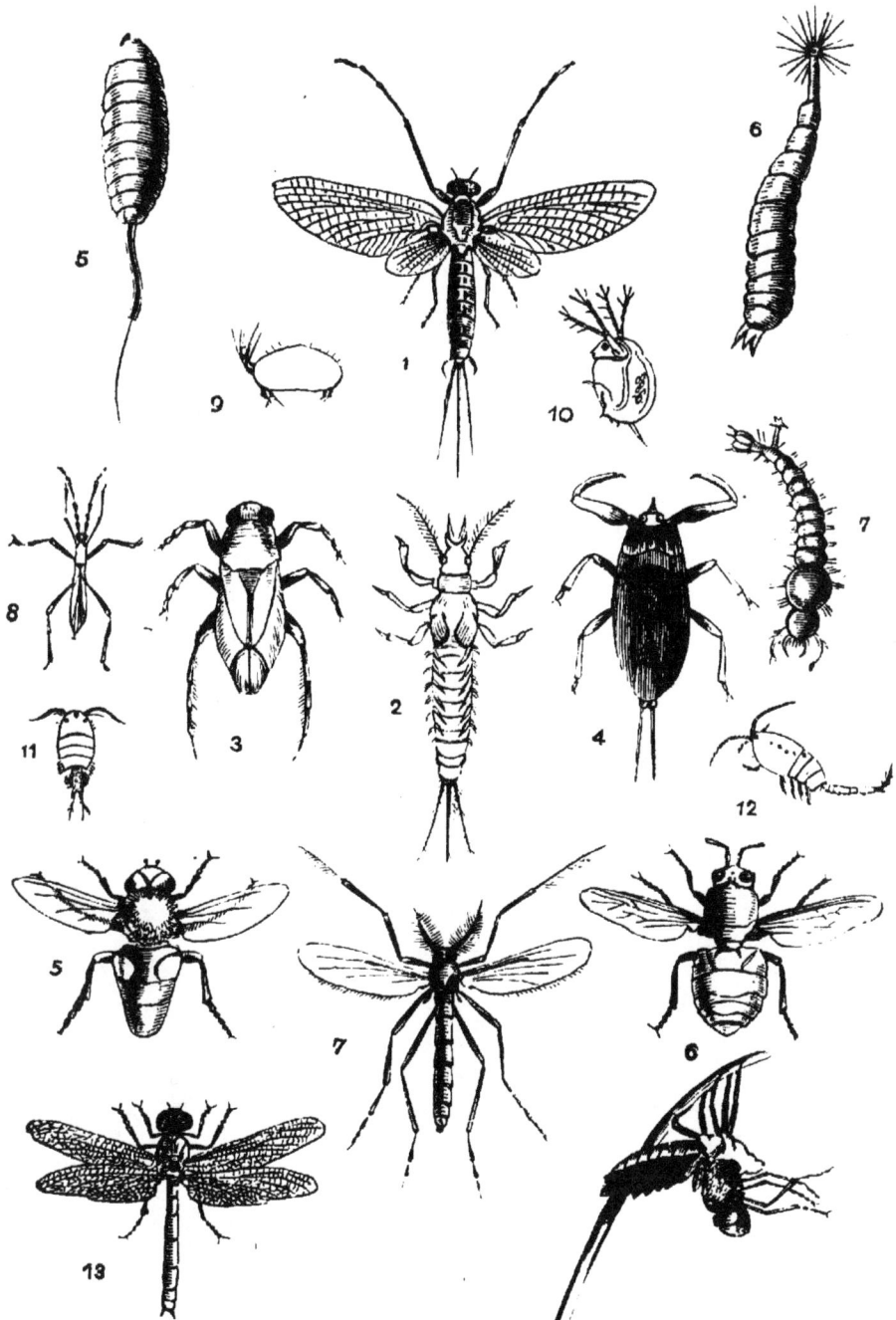

Plate XII

ROUTLEDGE'S USEFUL LIBRARY.

In fcap. 8vo, cloth limp or illustrated boards, 1s. each. (Postage 2d.)

Ladies' and Gentlemen's Letter Writer, &c.

Home Book of Domestic Economy. ANNE BOWMAN.

Common Things of Every Day Life. By ditto.

Rundell's Domestic Cookery. Unabridged.

Tricks of Trade in Food and Physic.

Notes on Health : How to Preserve or Regain It. By W. T. COLEMAN.

Common Objects of the Microscope. Rev. J. G. WOOD.

One Thousand Hints for the Table.

How to Make Money. By FREEDLEY.

Infant Nursing. Mrs. PEDLEY.

Practical Housekeeping. Do.

A Manual of Weathercasts and Storm Prognostics.

Commercial Letter-Writer.

Ready-Made Speeches.

The Dinner Question. By TABITHA TICKLETOOTH.

The Book of Proverbs.

Two Thousand Familiar Quotations.

The Book of Phrases and Mottoes.

Five Hundred Abbreviations Made Intelligible.

How to Dress on £15 a Year as a Lady.

How to Economise like a Lady.

The Guide to London.

Tables and Chairs.

The Competitor's Manual for Spelling Bees.

Breakfast, Luncheon, and Tea. By MARION HARLAND.

How we Managed Without Servants.

Shilling Manual of Etiquette.

The Pleasures of House-Building. By J. FORD-MACKENZIE.

Plate Swimming. By R. H. WALLACE DUNLOP, C.B.

Knots, and How to Tie Them. By J. T. BURGESS.

Price 1s. 6d. each.

Landmarks of the History of England. By Rev. J. WHITE.
Landmarks of the History of Greece. By ditto.
The Gazetteer of Great Britain and Ireland.

POPULAR LAW BOOKS.

Price 1s. each. (Postage 2d.)

The Law of Landlord and Tenant, with an Appendix of useful Forms, Glossary of Law Terms, and New Stamp Act. By W. A. HOLDSWORTH.

The Law of Wills, Executors, and Administrators, with useful Forms. By W. A. HOLDSWORTH.

The New County Court Guide

The Education Act, including the Act of 1873. By W. A. HOLDSWORTH.

Master and Servant By ditto.

The Ballot Act. By ditto.

The Licensing Act. By ditto.

The Law of Bills, Cheques, and I.O.Us. By ditto.

Friendly Societies. By ditto.

PUBLISHED BY GEORGE ROUTLEDGE & SONS

ROUTLEDGE'S CHEAP COOKERY BOOKS.

Francatelli's Cookery. 6d.
Soyer's Cookery for the People. 1s.
Mrs. Rundell's Domestic Cookery. 1s.
—— Another Edition, 1s. 6d., cloth.
Breakfast, Luncheon, and Tea. 1s.
Buckmaster's Cookery Book. 2s. 6d.
See also HOUSEHOLD MANUALS.

ROUTLEDGE'S HOUSEHOLD MANUALS.

Including the "HUNDRED WAYS" Cookery Books.
Price 6d. each. (Postage 1d.)

The Cook's Own Book.
The Lady's Letter Writer.
The Gentleman's Letter Writer.
The Village Museum.
How to Cook Apples.
How to Cook Eggs.
How to Cook Rabbits.
Every-Day Blunders in Speaking
The Lovers' Letter Writer.
Cholera. By Dr. LANKESTER.
Home Nursing.
How to Make Soups.
How to Cook Onions.

Ready Remedies for Common Complaints.
How to Dress Salad.
How to Cook Game.
How to Make Cakes.
The Lady Housekeeper's Poultry Yard.
How to Cook Vegetables.
How to Make Pickles.
The Invalid's Cook
How to Stew, Hash, and Curry Cold Meat.
How to Make Puddings.

ROUTLEDGE'S SIXPENNY HANDBOOKS.

With Illustrations, in illustrated boarded covers. (Postage 1d.)

Swimming.
Gymnastics.
Chess, with Diagrams.
Whist.
Billiards and Bagatelle.
Draughts and Backgammon.
Cricket.
The Card Player.
Rowing and Sailing.
Riding and Driving.
Shooting.   [sword.
Archery, Fencing, and Broad-

Manly Exercises : Boxing, Running and Training.
Croquet.
Fishing.
Ball Games.
Conjuring.
Football.
Quoits and Bowls.
Skating.
Fireworks.
500 Riddles.
Dominoes.

PUBLISHED BY GEORGE *ROUTLEDGE* & SONS.

Printed in the USA
CPSIA information can be obtained
at www.ICGtesting.com
LVHW010335310524
781891LV00006B/262
9 781022 548626